Mathematical and Algorithmic Foundations of the Internet

Chapman & Hall/CRC
Applied Algorithms and Data Structures Series

Series Editor
Samir Khuller
University of Maryland

Aims and Scopes

The design and analysis of algorithms and data structures form the foundation of computer science. As current algorithms and data structures are improved and new methods are introduced, it becomes increasingly important to present the latest research and applications to professionals in the field.

This series aims to capture new developments and applications in the design and analysis of algorithms and data structures through the publication of a broad range of textbooks, reference works, and handbooks. The inclusion of concrete examples and applications is highly encouraged. The scope of the series includes, but is not limited to, titles in the areas of parallel algorithms, approximation algorithms, randomized algorithms, graph algorithms, search algorithms, machine learning algorithms, medical algorithms, data structures, graph structures, tree data structures, and other relevant topics that might be proposed by potential contributors.

Published Titles

A Practical Guide to Data Structures and Algorithms Using Java
Sally A. Goldman and Kenneth J. Goldman

Algorithms and Theory of Computation Handbook, Second Edition – Two Volume Set
Edited by Mikhail J. Atallah and Marina Blanton

Mathematical and Algorithmic Foundations of the Internet
Fabrizio Luccio and Linda Pagli, with Graham Steel

Chapman & Hall/CRC
Applied Algorithms and Data Structures Series

Mathematical and Algorithmic Foundations of the Internet

Fabrizio Luccio and Linda Pagli
with Graham Steel

CRC Press
Taylor & Francis Group
Boca Raton   London   New York

CRC Press is an imprint of the
Taylor & Francis Group an **informa** business

A CHAPMAN & HALL BOOK

Cover Image: Giulio, "Connected worlds," acrylic and charcoal on canvas, 2006.

CRC Press
Taylor & Francis Group
6000 Broken Sound Parkway NW, Suite 300
Boca Raton, FL 33487-2742

© 2012 by Taylor & Francis Group, LLC
CRC Press is an imprint of Taylor & Francis Group, an Informa business

No claim to original U.S. Government works

Printed in the United States of America on acid-free paper
Version Date: 20110510

International Standard Book Number: 978-1-4398-3138-0 (Hardback)

Library of Congress Cataloging-in-Publication Data

Luccio, Fabrizio, 1938-
    Mathematical and algorithmic foundations of the internet / Fabrizio Luccio, Linda Pagli, Graham Steel.
        p. cm. -- (Chapman & Hall/CRC Press applied algorithms and data structures series)
    Includes bibliographical references and index.
    ISBN 978-1-4398-3138-0 (hardback)
        1. Internet--Mathematical models. 2. World Wide Web--Mathematical models. I. Pagli, Linda. II. Steel, Graham, 1977- III. Title. IV. Series.

    TK5105.875.I57L835 2011
    004.67'80151--dc22   2011008025

Visit the Taylor & Francis Web site at
http://www.taylorandfrancis.com

and the CRC Press Web site at
http://www.crcpress.com

Contents

List of Figures

Preface

Designing and maintaining a computer network is a task for professionals, and even a superficial understanding of its operation requires specialist knowledge that only a few people possess. At the same time, many people use the Internet daily without realizing how complex the underlying operations are.

This book is intended to partially fill this gap. It contains an introduction to the wealth of mathematical and algorithmic concepts and methods on which computer networks rely, and to some of their applications to the Internet and the Web. At first glance the list of chapters may appear slightly odd, but in fact they run from fundamental concepts towards more specific topics and applications. The mathematical treatment is rigorous, but the text is kept at a level adequate to readers with an elementary mathematical background. In any case some more technical parts may be skipped without preventing a general understanding of the text.

The book is intended for use in Computer Science courses at elementary level; or as a suggested reading for students in other fields; or for providing supplementary notions to technical professionals; or, finally, for curious people interested in the advancement of science and technology. Mathematical and algorithmic concepts and methods are accompanied by notions coming from literature, history, art, and other fields, to provide a lighter reading experience and to show the universality of many of the concepts treated.

Fabrizio Luccio and Linda Pagli are the main authors. Graham Steel provided literary assistance.

About the Authors

Born in Tripoli, **Fabrizio Luccio** holds a doctoral degree in electrical engineering from the Politecnico di Milano. He is currently a professor of computer science at the University of Pisa.

Previously he taught and conducted research in various fields of computing at the Massachusetts Institute of Technology (Project MAC), the University of Southern California, and New York University. After a move to Pisa, he spent several sabbatical periods abroad in leading academic institutions and industrial research centers. He also continues to work with UNESCO for the dissemination of informatics in developing countries. His major scientific interests are in computing systems and algorithmics.

Professor Luccio received several scientific and academic awards. He is a Life Fellow of the IEEE and a member of the ACM.

Born in Livorno, **Linda Pagli** holds a "Laurea" degree in information sciences from the University of Pisa and is currently a professor of computer science at the same university.

After a career as a researcher in computing, she was appointed a professor at the University of Salerno, returning to Pisa three years later.

Professor Pagli has been a visiting scientist at Carleton University in Ottawa and at Columbia University. She has pursued intense activities in higher education in favor of developing countries and has been a visiting professor at the Universities of Somalia and Botswana. Her current research interests are in the bases of computation and in design and analysis of sequential and distributed algorithms.

Professor Pagli is a member of the ACM.

Born in south London, **Graham Steel** holds a degree in mathematics from the University of Cambridge and a PhD in informatics from the University of Edinburgh. He has held positions at Universities in Germany and Italy and now works for the French National Computer Science Research Agency (INRIA) at the Specification and Verification Laboratory (LSV) of the École Normale Supérieure de Cachan, near Paris. He specializes in the security analysis of cryptographic modules used in devices such as cash machines and smartcards. He also writes light verse.

Chapter 1

An unconventional introduction to the Internet

How Euler answered a perplexing question about a journey in the historic city of Königsberg, and how that journey can be re-shaped for more interesting purposes using Internet mathematics.

The sketch map in Figure 1.1 is one of the most sacred images in the history of mathematics. Drawn by Swiss scientist Leonhard Paul Euler for an article published in 1736, it was intended to depict the connections among the different districts of the city of Königsberg via seven bridges crossing the river Pregel. Why Königsberg; why Euler; and, above all, why are we discussing this here?

For many centuries Königsberg, in the heart of Eastern Prussia, was one of the most important cities in the world. It deserved a better fate. Conquered by manifold rulers, invaded by various armies, the city finally became part of the Soviet Union under the name of Kaliningrad at the end of World War II, during which brutal aerial bombardment had destroyed almost all of its historic monuments. Even the celebrated university *Albertina*, founded in the sixteenth century by Albert of Brandenburg, fell victim to the bombs. It was here that Immanuel Kant spent his entire life; and, according to a slightly dubious tradition, it was also the origin of a tantalizing problem, finally brought to the attention of Euler in 1735.

At that time, Euler held a position in the Imperial Russian Academy of Sciences founded by czar Peter the Great in St. Petersburg. The problem posed to him was very simple, and scarcely interesting at first glance. Königsberg had seven bridges connecting four districts of the city separated by the bends of the river. In Euler's drawing the districts are indicated by A, B, C, D and the bridges by a, b, c, d, e, f, g. The historic center of the city was in A, the Kneiphof island, site of the University Albertina and (one century later) of Kant's grave. The question raised by the curious citizens was whether it was possible to take a walk starting from an arbitrary point of the city, crossing each bridge exactly once, and returning to the starting point. If you have not seen this problem before, consider spending a few moments on it. It should

1

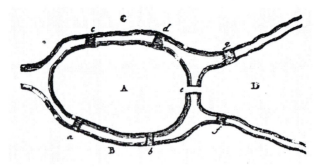

FIGURE 1.1: The city of Königsberg in the Euler's drawing.

not be difficult to convince yourself informally that no such a path exists, although proving this formally may be a little more difficult. So what could have raised Euler's interest?

The answer is fascinating, because what Euler extracted from the "seven bridges problem" was far more than the solution to a popular puzzle. His paper entitled *Solutio problematis ad geometriam situs pertinentis* (On the solution of a problem relating to the geometry of position: in the eighteenth century the language of science was Latin) marks the birth of what is now called *graph theory*, a branch of mathematics that the reader probably encountered at high school. After describing the popular problem on the bridges of Königsberg, Euler presented his work with the following words:

On the basis of the above I formulated the following very general problem for myself: given any configuration of the river and the branches into which it may divide, as well as any number of bridges, to determine whether or not it is possible to cross each bridge exactly once.

A *graph* is a mathematical object consisting of a set of *nodes* and a set of *arcs* showing relations between nodes. Graphically nodes and arcs are represented with dots and with lines joining dots in pairs, so the graph of Königsberg could be drawn as indicated in Figure 1.2(a), although a similar picture was never used by Euler. The number of arcs touching, or *incident* to a node is called the *degree* of that node, so for example node A has degree five. An alternative representation of the graph is given in Figure 1.2(b) with *weights* (number of bridges) associated to each arc.[1] A journey across all the

---

[1]In the language of mathematics, Figure 1.2(a) represents a *multigraph*, i.e., a graph with multiple arcs between pairs of nodes, and Figure 1.2(b) represents a *weighted graph*. Here a weight accounts for a number of connections, but the concept and use of weights is much more general. Directed arcs as in Figure 1.2(c) will be discussed below. Nodes and arcs are also called *vertices* and *edges*.

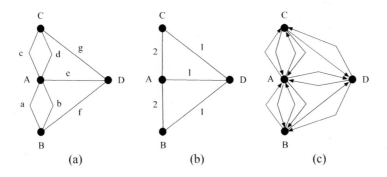

FIGURE 1.2: The graph of Königsberg: (a) with multiple arcs; (b)with weighted arcs; or (c) with directed arcs.

bridges without crossing any one of them more than once, if it exists, is called a *Euler tour* of the graph.

Today pedestrians in Königsberg are rare, and rather than reflecting on appealing mathematical problems while strolling around their thoughts are more likely to be angrily directed at reckless drivers that jeopardize their safety or splash their clothes with filthy water from the puddles. But we like to believe that in Euler's time the people of Königsberg were pleased to see their seven bridges problem solved so elegantly, although Euler's approach referred to a general arrangement of districts and bridges, using the particular problem of Königsberg only as a starting point for the creation of a new theory.

In the realm of graphs, Euler's solution can be stated as follows:

A graph admits a Euler tour if and only if all its vertices have even degree.

If you have tried to solve the Königsberg problem yourself, you might have conceived a condition equivalent to the *only if part* of Euler's theorem. In fact even if a single node has odd degree a tour cannot reach and then leave it, one or more times, without either traversing an incident arc twice or leaving one incident arc untraversed. In the case of Königsberg, for example, node B has odd degree three (in fact, all the districts of the city have an odd number of bridges) so the walk is impossible. To convince yourself of Euler's assertion note that after entering district B from bridge a one can leave B from bridge b, enter again from bridge f, but at this point it would be impossible to leave B without crossing one of the three bridges again.

More difficult is to prove the *if part* of the theorem, i.e., if all the vertices have even degree a Euler tour exists. We leave the proof of this part to any graph theory textbook. Even so, a reader with mathematical skills can at least try to prove it without any help. In particular a reader familiar with computer algorithms may produce a *constructive proof*, i.e., the scheme of a program that builds a Euler tour for any graph where all the vertices have even degree. We will come back to this in a next chapter. It is time, instead, to direct our attention to a different type of journey inside Königsberg that has much more to do with computer networks. But first, a crucial observation.

This book is about the Internet and the Web, two completely distinct but strictly related entities. Although these entities will be discussed thoroughly in later chapters, we assume the readers to be familiar with their basic characteristics, elementary use, and most common terminology. What is important, however, is to understand that both the Internet and the Web can be studied in form of graphs, with a substantial difference between them. The Internet is a *physical network* composed of computers (the nodes) connected to one another via cables, optical fibers, wireless connections etc. (the arcs). This definition is rather imprecise as will become clear in the following, but it is nevertheless sufficient, as a first approximation, to treat the Internet as a graph as we did with the city of Königsberg. The Web, or better the World Wide Web or www, is an *abstract network* composed of Web pages (the nodes) connected via clickable links (the arcs).

Two basic features characterize the two networks when treated as graphs. Like the bridges of Königsberg, the physical data lines of the Internet can generally be traversed in both directions. The links of the Web, however, have an orientation from a pointing page to a pointed page, and can be traversed only in one direction. We then speak of a *directed graph* whose arcs are represented as arrows with the butt in the pointing node and the head in the pointed node. An undirected graph can always be transformed into an equivalent directed graph by substituting each arc of the former with two arcs pointing in opposite directions, as shown in Figure 1.2(c). An inverse transformation is clearly possible only if both the directed arcs exist.

The second important feature of the graphs of the Internet and of the Web is that, in principle, there is no correspondence between the nodes of the two networks, or between the arcs of them. One computer (node of the Internet) can host several Web pages (nodes of the Web). Conversely, one Web page can be replicated in several computers. The arcs are also unrelated, since two Web pages connected by a link may reside in two computers that are not connected by a line, or two pages hosted in the same computer may share a link that has no corresponding arc in the Internet. Conversely, two computers connected by a line need not host linked pages. Amazingly we shall see that, in spite of theses distinctions, some mathematical laws governing both networks are the same.

Now go back to Königsberg, find a friendly looking pub, and sample some of the local beer. Then, as you may find you have consumed a little too much of it, make a journey through the city without any particular destination or constraint. Forgetting Euler you are now free to cross any bridge however many times you like, or never cross it at all. Meanwhile we will monitor your steps on a city map. As long as you remain in one district you will not be in any danger, so it suffices for us to use a map drawn in graph form (for example the one of Figure 1.2(a)): if you do not leave a district, you stay in a single node of the map. However drunkards are invariably attracted by water, so sooner or later you will cross a bridge chosen at random, thereby changing node in our map. For us, your tour is a *random walk in a graph*, where italics indicate an expression of the mathematical jargon.

Although this may be surprising, we will show that this way of looking at things is a basic constituent of Web search engine techniques, where the mathematical properties of random walks have been exploited with great success to increase the relevance of the answers that we obtain from the Internet. So be prepared to face some non trivial issues in probability theory. For the moment we merely anticipate some facts regarding your journey in Königsberg, according to a probabilistic process known as a *Markov chain*. In particular we wish to inquire if you have a chance of eventually reaching your hotel.

From any district, assume you take any of its bridges with equal probability. If you walk long enough you will be passing through all districts, although with different frequencies depending on the overall connectivity of the map. A neat result could be drawn as a limit of the mathematical process if your journey lasted forever. However as human strength has a limit you will eventually stop after a finite number of steps, so we will only be able to make a heuristic guess of the district where you end up. The longer your journey has been, the more accurate our estimate will be. An interesting part of the story is that, for an infinite journey, the probabilities of passing through each district are independent of the place from which you start, so our estimate will be reasonably correct if you have walked enough.[2]

If all this is applied to the map of Königsberg it can be proved that you will end up in node A with probability roughly double than of any other node. This is not surprising since Kneiphof has more connecting bridges than any other district. However making predictions without theoretical support is dangerous, particularly for random walks that depend on *all* the arcs of the graph. In fact, while the probability of ending in B and C is the same because the two nodes are connected to all the others exactly in the same way, it is not at all obvious that the probabilities of ending in any one of them, or in D, are very close to each other as in fact the case. When studying the Web graph we will discover that these probabilities are among the most popular parameters for deciding the ordering in which the pages are returned in response to an Internet query.

Another indicator of relevance is based on a very different concept. Around noon the next day, after a good sleep, you decide to go looking for art work to buy. Since the index of your guidebook mentions one art museum in district B, you conclude that there may be some art galleries in the surrounding area. So in addition to B you decide to visit the *adjacent* districts A and D (i.e., the nodes connected to B in the graph), but not C that is far (i.e., two bridges)

---

[2] As you may imagine, these properties are strictly true only if some precise mathematical conditions are satisfied. We will go back on this in a later chapter, and take the present description as an informal preview.

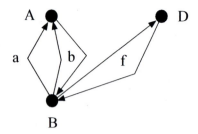

FIGURE 1.3: A sub-graph of Königsberg possibly relevant for buying art work.

away. Moreover you hear in the news that, for maintenance reasons, bridge e is temporarily closed, and bridge a can be traversed only in the direction from B to A. So essentially you limit your walk to a portion of the map corresponding to the directed *sub-graph* of Figure 1.3. Now you have to decide which of the three districts is likely to be the most useful for you, and start looking for art work from there.

To rank the districts, once again we invoke some ideas born in the realm of Web search engines. Looking for art work to buy close to an art museum may be a good rule of thumb, but the best galleries are not necessarily in those districts. However, since one can reasonably visit only a smallish portion of the graph, the walk is extended only to the adjacent districts. But how can we rank the relevance of these? One criterion is deciding, among the selected districts, which ones will probably host the best galleries (called *authorities* in the Web jargon), and which ones are good starting points for the visit (called *hubs*). Note that in a directed graph the two sets are generally distinct because good hubs will contain several outgoing links to the most authoritative sites, and strong authorities will have several incoming links from the best hubs.

The computation involved is not elementary and will be discussed in a following chapter. If applied to the graph of Figure 1.3 it indicates that the most authoritative district is A and the best hub is B. With a little thought this result is not surprising. Not only is B the district where the art museum is located, but it is also the only one with outgoing links to the other two districts; hence it is a good starting point for reaching all authoritative districts directly. On the other hand A has two incoming links from the best hub B, meaning that it is a favorite place to visit in the area, and so some major art galleries have probably been established there. Just as for random walks, these results depend solely on the structure of the graph.

We conclude this informal introduction by emphasizing that anyone who really wants to understand how the Internet and the Web are organized and function must be ready to study a fair bit of mathematics and computation theory.[3] In fact, few technological fields are so tightly connected with hard sciences as computer networks are. In the next chapters we will start discussing discrete mathematics and computational complexity, linear algebra and probability, with some incursions in the world of numbers. Only then we will put all these pieces of knowledge together, to explain how computer networks work. Since there are too many mathematical prerequisites to treat them all together, we introduce them in separate chapters, going in to some depth for each one of them, but keeping the explanations as accessible as possible. We will show that many mathematical concepts have a counterpart in everyday life and appear in the most unexpected situations.

Bibliographic notes

Euler's original paper on the seven bridges of Königsberg, written in 1736 but not published until 1741, can be read translated in English in the excellent set of books: Newman, J.R., Editor. 1988. *The World of Mathematics*, Vol. 1. Tempus Books, recently reprinted. The interest may only be historical, but it is impressive to discover how a great mathematician gave rise to a new theory with such extreme simplicity. Scholarly notes on the birth of Euler's paper can be found e.g. in: The Truth about Königsberg, by B. Hopkins and R. J. Wilson, http://www.nku.edu/ longa/classes/mat385/ resources/Hopkins1EulerKonigsberg.pdf

For a general introduction to networking one can start from: Barabási, A.L. 2002. *Linked: The New Science of Networks*. Perseus Publishing, Cambridge, MA. This has been the first book on the subject directed to a general public. Reading it requires a modest knowledge of mathematics, but is still recommended for the excellent organization of the text. The author is a major international expert in the field.

Other books, papers, and Web sites that may constitute a useful complement to this book will be indicated in each specific chapter.

---

[3]Of course the ones interested in hardware also have to study physics. This aspect, however, is outside the scope of this book.

Chapter 2

Exponential growth

How some things rocket to an incredible size even though their growth may be hard to observe in its initial steps.

The rich canon of humorous prose by Mark Twain includes a satire of the "magnanimous-incident" anecdotes of the Victorian age, in which the author claims to have investigated the true sequel of several such stories. A stray poodle with a broken leg that, after having been healed by a benevolent physician, shows up gratefully the next day in the company of another stray dog requiring acute limb treatment, is the seed of a logical chain of events in which the number of dogs waiting in front of the physicians door doubles every day to become four, eight, etc., inspiring new sentiments in the physician's mind:

This day also passed, and another morning came; and now sixteen dogs, eight of them newly crippled, occupied the sidewalk, and the people were going around. By noon the broken legs were all set, but the pious wonder in the good physician's breast was beginning to get mixed with involuntary profanity. The sun rose once more, and exhibited thirty-two dogs, sixteen of them with broken legs, occupying the sidewalk and half of the street; the human spectators took up the rest of the room. The cries of the wounded, the songs of the healed brutes, and the comments of the onlooking citizens made great and inspiring cheer, but traffic was interrupted in that street. The good physician hired a couple of assistant surgeons and got through his benevolent work before dark, first taking the precaution to cancel his church-membership, so that he might express himself with the latitude which the case required.

But some things have their limits. When once more the morning dawned, and the good physician looked out upon a massed and far-reaching multitude of clamorous and beseeching dogs .... [1]

....he realized that the thing had gone along far enough, set out to resolve

---

[1] Mark Twain, *The Grateful Poodle*, in *About Magnanimous-Incident Literature*.

it with a shotgun, got bitten by the original poodle, and passed away in the horrendous convulsions of hydrophobia.

The evolution of the set of dogs follows an *exponential growth* in the genuine (i.e., strictly mathematical) sense of the phrase. The expression has become quite popular in everyday language to evoke any process that grows at very high speed, without the implication of making any actual measurement of it. There is nothing to object to about this informal usage, apart from perhaps the fact that it misses a subtle effect of exponential growth that is inherent to the mathematical definition. In fact, if the rate of growth is constant but small the phenomenon is hard to detect in its initial steps, until it grows so fast as to become clearly evident and to overtake any competing non-exponential process from this point on. So, let us start by recalling how exponential growth is defined.

A mathematical function $f(t)$ grows exponentially with t if its next value is increased over its current value of an amount proportional to this current value. Very often, but not necessarily, t represents time (hence the letter chosen here for this variable), and the growth is applied at discrete intervals. For unitary intervals we have:

$$f(t+1) - f(t) = Kf(t), \text{ or equivalently } f(t+1) = Gf(t) \qquad (2.1)$$

where $K > 0$ is a constant, and $G = K + 1 > 1$ is the *growth factor*.[2] Iterating formula (2.1) we have the familiar expression:

$$f(t) = AG^t \qquad (2.2)$$

where $A = f(0)$ is the value of the function at the initial time $t = 0$. The independent variable t appears as an exponent in the expression (2.2), accounting for the name exponential growth. In this expression the critical term is G^t, not A, no matter how big this latter value may be. If we wait long enough the value of G^t explodes.

An immediate application of formula (2.2) is to compute the amount of money deposited in a savings account, where the interest is compounded once per year. Even a generous interest rate causes a small growth in the first few years, but in the long term this would be by far the best investment if inflation or bankruptcy did not spoil the game and one lived long enough to cash the balance. For example starting with a deposit A of one dollar, an interest rate of $K = 5$ percent, i.e., $G = 1.05$, would yield a total of $1.05 after one year, $1.10 after two years, $1.63 after ten years; but then $11.47 after fifty years, $131 after one hundred years, over two million dollars after three hundred years, over five thousand billion dollars after six hundred years. Nobody without mathematical skill could have predicted such growth from looking at the figures of the first ten years.

Although the definition of exponential growth refers to a parameter G that

---

[2]For $K < 0$ we have $G < 1$ and the function actually decreases with t, with an *exponential decay*.

is an arbitrary real number, the structures that we will use for network description and the computations that we will face are often confined to integers. So for us G is generally an integer greater than one. Let us start with $G = 2$ and the expression (2.2) in the form $f(t) = A2^t$, or $f(t)/A = 2^t$, a value that doubles at every step. Starting from time $t = 0$ we have $2^0 = 1$ and then the sequence 1, 2, 4, 8, 16 etc. Under the so called *Malthusian growth model* this is a possible scheme for world population growth, assuming that each couple of parents generates an average of four children. Here the power of mathematics is frightening, because in the sequence generated by 2^t each element is greater than the sum of all the preceding ones (e.g., $16 > 8 + 4 + 2 + 1$).

As we shall see in a later chapter this property has unexpected importance in the study of random processes, but in reference to the population of the globe, it simply indicates that there are more living beings today than in the whole history of mankind, even if we count only the latest generation (i.e., we assume that parents, grandparents etc. have all died).[3] Passing from the human to the bacterial realm this growth model applies almost exactly to the number of microorganisms present in a biological culture where each organism splits into two new ones so as to double the total number of organisms in any generation, typically every twenty minutes or so.

It is worth knowing that the same mechanism was applied to a revolutionary technique known as PCR for *Polymerase Chain Reaction* that had an enormous impact in molecular biology. It was conceived by Kary Banks Mullis, an amazing scientist and surfer from Newport Beach, California, who won a Nobel Prize in chemistry for the invention. Without going into any detail, PCR allows a virtually endless multiplication of identical DNA strands needed in genetic experiments by making two copies of each existing strand at each phase. In this way millions of copies of a strand are produced in a very short time. When the technique was disclosed in 1985, many scientists were surprised that no one had previously thought of applying the power of exponential growth to molecule production. Today PCR affects our everyday lives because it is commonly used to carry out a wealth of lab activities such as diagnosing genetic diseases, detecting viruses, and so on.

Aside from science, this law of doubling has been rediscovered many times in popular history. A well known chess myth tells how the game was invented by an Indian Brahmin named Sissa to satisfy a request of the Rajah Balhait. When asked to name his reward, Sissa replied that he would content himself with an amount of wheat to be determined by putting one (i.e., $2^0 = 1$) grain on the first square of the chessboard, and doubling the number of grains on each of the following squares up to last one. The rajah was amazed by the modesty of the request, but it turned out that all the barns in his kingdom were not sufficient to supply the grains required to reach the astronomical number of 2^{63} on the 64-th square: a number that requires twenty decimal digits to

---

[3]Incidentally this requires an updating to the mechanism of metempsychosis, and it is amazing that this belief is mostly accepted by peoples that multiply at a frantic pace.

write down. And if it was natural for the Indians to include arithmetic in their tales, it may be surprising that an ancient Christian author pointed out that if a true believer had converted a heretic in one year, and then the two had done the same with two heretics in the next year, and so forth, more than one million heretics would have been converted in just twenty years.[4]

In the Internet era, computer scientists more than holy men are concerned with the phenomenon of exponential growth, because solving some important problems may require an exponential number of operations and the running time of the corresponding computer programs becomes enormous even on small input data. In this respect it is not surprising that Kary Mullis, the aforementioned inventor of PCR, said that he was familiar with the phenomenon of exponential growth that is at the base of his technique because he "had been spending a lot of time writing computer programs." Let us introduce this way of seeing things by taking a new walk through the bridges of Königsberg encountered in Chapter 1, and postpone a deeper study of exponential *computation* to a subsequent chapter on algorithms.

If you have devoted some attention to the problem of finding a Euler tour in a graph, you may have asked yourself how it was possible that the people in Königsberg did not realize immediately that such a tour could not be taken in their town. Presenting the problem in the foreword of his paper, Euler adds:

I was told that while some deny the possibility of doing this and others were in doubt, there were none who maintained that it was actually possible.

That is, nobody exhibited a solution (and we know one does not exist), but many were in doubt. And Euler himself accepted this doubt as being natural. The question is: how could the citizens of Königsberg *prove* that a tour was impossible? The discovery of Euler's theorem provided a criterion for taking a decision in a short time, because it stated that it suffices to count the number of bridges leading to each district of the city. If all these numbers are even a tour exists, otherwise it does not. But, before that, the only way to find out was to attempt all possible tours crossing seven bridges and check whether one of them really did cross each bridge exactly once. Such a proof method, called *by brute force*, is legitimate from a logical point of view but in general is computationally unfeasible, as in our case where, as we will now see, the number of possible tours is an exponential function of the number of bridges.

The simplest way of enumerating the tours is to build all the *permutations* of the bridge names a, b, c, d, e, f, g, that is all the possible arrangements of these names in a sequence, and test whether one of them corresponds to a tour in the graph of Königsberg reported in Figure 1.2(a) of Chapter 1. The first permutation $a\,b\,c\,d\,e\,f\,g$ can be followed on the graph starting from node A and taking the arcs $a\,b\,c\,d\,e\,f$ one after the other until node B is reached. The attempt ends here, since the following bridge g in the permutation does not

---

[4]The holy man thought he had proved the power of faith but had actually proved a "power of two" (as $2^{20} > 1,0000,000$).

touch B. Then a new permutation is tested, e.g., $a\,b\,c\,d\,e\,g\,f$, and this attempt ends in node C after arc g has been followed. Of course we need an algorithm to enumerate all the permutations, but this a different story.[5] What we would like to compute first is how many such permutations there are.

From elementary combinatorics we know that the number of permutations of n elements is given by the *factorial* of n denoted by $n!$, i.e., the value $n! = 2 \cdot 3 \cdot 4 \cdot .... \cdot n$.[6] Probably everybody knows that this function grows very fast with n, and in fact for the seven bridges problem we have $7! = 5040$. No wonder that the citizens of Königsberg did not have the patience to try them all.[7] To compute the value of $n!$ without having to perform the $n - 2$ multiplications that appear in its definition, we can make use of the identity:

$$\lim_{n \to \infty} \frac{\sqrt{2\pi n}\,(n/e)^n}{n!} = 1$$

from which the well known approximation can be derived:

$$n! \sim \sqrt{2\pi n}\left(\frac{n}{e}\right)^n.$$

This formula, known as Stirling approximation, gives an increasingly accurate value of $n!$ for increasing n, and in particular shows that the function grows exponentially with n (and grows a lot since $e = 2.718...$ is a constant, then in the term $(n/e)^n$ both the base n/e and the exponent increase linearly with n). For example $30! > 2 \cdot 10^{32}$.

Now let us take a look at the center of Kaliningrad, the former Königsberg, as it is today. Two of the historic bridges are gone, and a new bridge h has been built directly between districts B and C. The new graph is indicated in Figure 2.1(a): unfortunately all the nodes maintain an odd degree and the citizens still cannot take a Euler tour. So we have proposed to the Russian municipality that they build two new bridges x, y as indicated in Figure 2.1(b), to finally fulfill the peoples expectations. In so doing all the nodes would have an even degree and a nice Euler stroll would be possible. But how can we discover its course? Or better, how can we determine a Euler tour in an arbitrary graph with even degrees? Note that finding such a tour amounts to proving Euler's theorem.

---

[5] Enumerating all the permutations of a set of n elements is an elementary but not trivial task. An elegant way is listing the permutations recursively, starting with the ones beginning with the first element (a in the example), then with the second element, etc., followed by the permutations of the other $n - 1$ elements listed recursively with the same criterion. Readers with algorithmic skills might attempt to write a recursive computer program for doing this.

[6] This result, known to the Hindus in the 12th century, can be seen as an immediate consequence of the recursive definition of permutations indicated in the previous footnote.

[7] With some ingenuity, the permutations for the problem at hand could be divided into groups each containing n cyclic shifts of the same permutation, e.g., $a\,b\,c\,d\,e\,f\,g$ - $b\,c\,d\,e\,f\,g\,a$ - $.... $ - $g\,a\,b\,c\,d\,e\,f$. In fact all the permutations of a group are equivalent for testing the existence of a Euler tour, so the number of trials could be lowered from $n!$ to $n!/n = (n - 1)!$. In the case of Königsberg this number becomes 720.

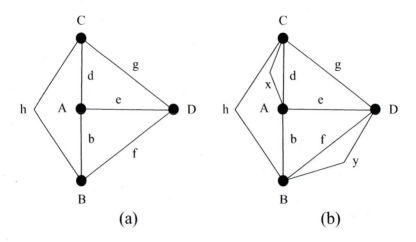

FIGURE 2.1: (a) The graph of Königsberg today. (b) The same graph with two more arcs x, y for making a Euler tour possible. Starting in B, one such a tour is $y\,f\,b\,e\,g\,x\,d\,h$.

As algorithms will be discussed in a subsequent chapter, we invite the reader to try solving this problem informally as follows. Go around in the graph until a node is reached twice, thus determining a cycle. Delete the arcs of this cycle from the graph. Take another walk until a second cycle is built, delete its arcs and continue in this way until no more arcs remain. You will end up with a set of cycles that touch each other in some nodes where the cycles can be merged in pairs to finally form a unique tour. In Figure 2.1(b) you can take a first walk starting from node C and following the arcs h, y, f: at this point you have touched node B twice determining the cycle $y\,f$. Delete these two arcs from the graph and try another walk until another cycle is formed, for instance $h\,b\,e\,g$. Delete these arcs and you are left with the last cycle $x\,d$. Now cycles $y\,f$ and $h\,b\,e\,g$ meet in B and are merged as a unique cycle $y\,f\,b\,e\,g\,h$, in turn merged with $x\,d$ in C to form the final Euler tour $y\,f\,b\,e\,g\,x\,d\,h$ that begins and ends in B. Proving that a graph with all even degrees admits such a construction should not be that difficult.

Euler tours appear in diverse algorithms related to distributed computation, together with tours of another type that are logically very closely related, but computationally very different. When some computers of a network are called upon to carry on a cooperative action, several operations are required to set up their connections and distribute the work among them. Although all this will form the bulk of a subsequent chapter on distributed computing, we consider at the moment the "simple" problem of choosing a set of communication lines to form a *ring* of computers. That is, in the graph of all existing connections we have to select a subset of arcs forming a cycle that traverses

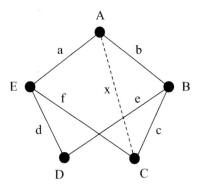

FIGURE 2.2: The Hamiltonian cycle problem.

each *node* exactly once. This is a sort of complementary problem compared to the one of Euler, where nodes are considered instead of arcs. In Königsberg it amounts to finding a walk that touches all districts exactly once and comes back to its starting point, as for example the elementary sequence *A C D B A* would indicate. Note that this may imply not crossing some of the bridges. However it turns out that the new problem is in general far more difficult than finding a Euler tour.

The new walk that we are interested in is called a *Hamiltonian tour* (or *cycle*) in honor of the Irish scientist Sir William Rowan Hamilton who invented a mathematical puzzle in 1857 called the *Icosian game*, that asks the solver to find a path traveling along the edges of a dodecahedron made of wood, such that the vertices are touched exactly once. Clearly those vertices and edges were nodes and arcs of a corresponding graph. Since the map of Königsberg is too simple for studying the new problem, consider the new graph of Figure 2.2. With some simple trials the reader may verify that if the graph is limited to the solid arcs a, b, c, d, e, f no Hamiltonian cycle exists, while with the addition of edge x the cycle *A B D E C A* appears. But what can we say in general for an arbitrary graph?

The present status of computational complexity theory rules out the possibility of finding a simple criterion applicable to any graph to decide whether a Hamiltonian cycle exists or not, as is possible for a Euler tour. The adjective "simple" has a technical meaning here. It is attached to functions that can be calculated in a "short time" as it will be mathematically specified later in the book. So we have to revert to a brute force approach, testing all the permutations of the nodes to determine whether one of them corresponds to a Hamiltonian cycle.

For the graph of Figure 2.2 (solid arcs) we have $5! = 120$ permutations of the nodes.[8] Namely:

$$ABCDE \quad ABCED \quad ABDCE \quad ABDEC \quad ABECD \ ..... \ EDCBA$$

In the first permutation $ABCDE$ we follow the path bc and stop here because there is not a next arc between the pair of nodes C, D. In the second permutation $ABCED$ the path $bcfd$ can be followed up the last node D, but we cannot return to the starting vertex A because there is no arc between D and A. Then this permutation does not correspond to a Hamiltonian cycle either. Similarly, all the other permutations do not correspond to a Hamiltonian cycle, that in fact does not exist in this graph. Note however that, if arc x is added, the forth permutation $ABDEC$ gives a solution because the starting node A can now be reached from C.

Can we then rely on such a brute force method, as no other essentially more efficient technique is known? For a small number of nodes, that is for not too many computers to be connected in a ring, we can patiently examine all permutations possibly with the aid of a computer program. But assume that the nodes are just thirty. We have already seen that $30! > 2 \cdot 10^{32}$. Assume that our computer can examine one permutation per nanosecond (10^{-9} sec, a hypothesis well out of the reach of present day hardware). In one second, 10^9 (one billion) permutations would be examined. However since there are "only" 31,536,000 (about $3 \cdot 10^7$) seconds in a year, we could examine about $3 \cdot 10^{16}$ permutations per year from among the more than $2 \cdot 10^{32}$ such permutations. We conclude that about $2 \cdot 10^{32}/(3 \cdot 10^{16}) > 10^{15}$ years, that is more than ten thousand billions centuries, will be necessary to complete the job for only thirty nodes. This shows the destructive strength of exponential growth. As we shall see in a next chapter, probability theory will help us to survive.

After having defined the phenomenon of exponential growth and discussed how its effects appear in different situations and for different values of the parameters involved, let us finally explain why this subject appears right at the beginning of the book. Although mainly mathematical, the theory on which computer networks logically rely has much to do with combinatorial and computational analysis. While working on them on the next chapters, the ghost of exponential growth will appear so frequently that a clear knowledge of it is an inescapable prerequisite. And after all, one has to start somewhere.

Bibliographic notes

To browse through a huge number of combinatorial formulae one of the best references is the book: Knuth, D.E. 1968. *The Art of Computer Programming.*

---

[8]We list these permutations as indicated in footnote 6. For finding a Hamiltonian cycle, they could be divided in groups each containing all the cyclic shifts of the same permutation as indicated in footnote 8 for Euler tours. For n nodes this amounts to dividing the total number of trials by n, a negligible improvement as shown.

Vol.1: *Fundamental Algorithms*. Addison-Wesley Publishing, Reading, MA. (In particular see the mathematical development leading to Stirling's formula.) We will refer several times to this fundamental work.

As a reference taken from the general literature, we note that Mark Twain's *About Magnanimous-Incident Literature* is not easily found in print but is largely available on the Web. It is a very amusing reading although (or perhaps because) it has no relation to the Internet.

To learn something about PCR in molecular biology without going too far from the world of computers, refer for example to the book: Setubal, J. and J. Meidanis. 1997. *Introduction to Computational Molecular Biology.* PWS Publishing, Boston, MA.

Chapter 3

Sequences and trees

How from the I Ching to the Internet era a wealth of codes have been developed for expressing information in many forms.

Throughout history, oracles have been consulted to obtain advice on important decisions, even in everyday life. The most ancient answers were binary, limited to "yes" and "no"; as in the *I Ching*, the Book of Changes, one of the oldest Chinese texts. Answers were represented by horizontal lines, a solid line for yes and a broken line for no.

It seems that the need for a more nuanced response emerged even in early times, so blocks of two lines were introduced giving rise to the $2^2 = 4$ of solid and broken lines combined. Later a third line was inserted in each block to double the possible combinations, and the $2^3 = 8$ basic signs of the I Ching, called *trigrams*, were born. These eight signs were representative of the eight elementary constituents of heaven and earth, and have survived with their whole expressive power. Look at them in Figure 3.1, arranged in their classical ordering.

The story, however, is just at the outset. The eight signs represent changing states, images perpetually mutating. The focus is not on the things in their being, but rather on their transitions. To obtain an even greater multiplicity the trigrams were combined in blocks of two, to form the complete set of

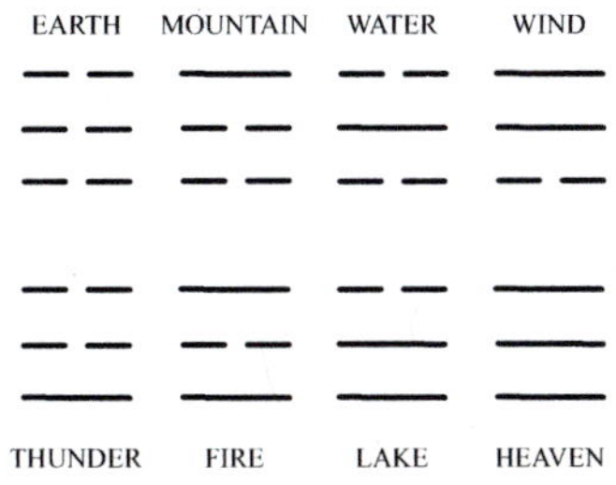

FIGURE 3.1: The I Ching trigrams.

the $2^6 = 64$ *hexagrams* contained in the book. Depending on the position in the hexagram, some of the lines may be changing, thereby increasing even further the number of possible oracular sentences. This wealth of cases is not surprising. The Book of Changes is a monument of Chinese wisdom, finding in the signs sufficient power to express itself.

Incrementing the number of lines, or considering a double possible meaning of each line, we can increase exponentially the number of oracular sentences to express a practically unlimited set of situations with a reasonably small set of lines. The ruling phenomenon is that of exponential growth, as discussed in Chapter 2, which actually regulates the field of communication making it all possible. Let us spend a little time on this point that seldom receives due attention.

3.1   The expressiveness of sequences

Human communication, either in sound or in images, develops through a sequence of elementary units. Writing is the natural starting point. Take any natural language and consider all the characters of its alphabet, that is, letters, numerical digits, punctuation marks, and other typographical characters including the blank space. The set of all possible sequences of characters is infinite if we do not bound their length, although a lot of them do not belong to the language. The number of characters used in most modern languages, with the important exception of Chinese and Japanese, is limited to a few dozen, but if they used a hundred, or a thousand, or just two, only the length of the words would change without affecting the expressiveness of the language.

With $c \geq 2$ different characters we can form $N = c^n$ different sequences (or words, or sentences) of length n. This is a huge number, even for limited values of n, and less sensitive to the value of c than one may imagine. In fact inverting the above expression we have $n = log_c N$. To construct the same number N of different sequences with an alphabet of d characters instead of c, the length of the new sequences becomes $n' = log_d N$, and from elementary mathematics we have:

$$log_d N = log_d c \cdot log_c N \Rightarrow n' = log_d c \cdot n \qquad (3.1)$$

where $log_d c$ is a multiplicative constant that depends only on the two bases. For example using the English alphabet of $c = 26$ characters or another arbitrary alphabet of only $d = 10$ characters, the relation between the length of the sequences would be:

$$n' = log_{10} 26 \cdot n \sim 1.4 \, n. \qquad (3.2)$$

Put simply, in the second language, words are not much longer than in the first one.

However, a real difference, and a very negative one for people in a hurry, would arise with a language of only one character. Take $N = 7$ different concepts like the Seven Sisters, a cluster of stars known as the Pleiades, and try to express them with the single character *s*. Among the many possibilities one can choose the seven sequences: *s, ss, sss, ssss, sssss, ssssss, sssssss*, that is, the sign s repeated from one to seven times. Whatever representation is used there will be at least one of the concepts requiring at least seven *s*'s, that is a sequence whose length is at least equal to the total number of concepts to be expressed. If there are one thousand such concepts, the longest word would consist of at least one thousand *s*, and all words would have an average length of at least five hundred. Only a very primitive or maybe esoteric communication system could be organized in this way, requiring great patience to compose the words, a high level of concentration to distinguish between them (especially when talking ...), and limiting *de facto* the expressive power of the language. Unlike all real languages that exhibit an exponential growth in expressivity, the number of words that can be expressed with only one character grows linearly with their length, making their use unfeasible.

With sequences of characters we can express all the words of a language, and hence all the knowledge that has been expressed in this language. Not satisfied with just one language, we can acquire universal knowledge of all the texts written in all languages. Then we can go further, not limiting ourselves to the existing languages, but considering innumerable languages born from the imagination, for which we can create dictionaries, grammars, literary masterpieces, and not yet written novels. How far can we take this idea? The problem has excited many thinkers and was brought to its extreme conclusion by Jorge Luis Borges in his masterly description of the Library of Babel, a library that contained a huge and unknown number of mysterious volumes which nobody was ever able to understand, until a librarian of genius discovered the fundamental law. He realized that:

" .... *the library is total and that its shelves register all the possible combinations of the twenty-odd orthographical symbols (a number which, though extremely vast, is not infinite): Everything: the minutely detailed history of the future, the archangels' autobiographies, the faithful catalogues of the Library, thousands and thousands of false catalogues, the demonstration of the fallacy of those catalogues, the demonstration of the fallacy of the true catalogue, the Gnostic gospel of Basilides, the commentary on that gospel, the commentary on the commentary on that gospel, the true story of your death, the translation of every book in all languages, the interpolations of every book in all books.*"[1]

---

[1]Borges J.L. 1962. *The Library of Babel*, in: *Labyrinths, Selected Stories and Other Writings*. New Direction Publishing. Translated from Spanish by J. E. Irby. Borges, a scholar and writer from Argentina, is a giant of 20th century world literature. Some of his ideas will be mentioned again in this book. Although Borges' writings have nothing to do with computation, they contain fantastic inventions that are often attractive to computer scientists.

Unfortunately the location of the books did not follow any understandable order, with terrible consequences that curious readers can discover in the original story.

Using sequences of characters, humans express all the words of a language, but computers express much more. In fact all information processed electronically is conveyed in the form of sequences, possibly flowing in parallel streams if many processors work together. For example, the visual information that we collect from a monitor or from a digital photograph is represented with a sequence of characters representing the features of the atomic portions of an image, as we shall see below.

A particular alphabet worth special consideration is the alphabet of numbers. Today all peoples use the same decimal notation to express numbers. Only the graphical shape of the ten digits may differ from one language to another, as for example in English, Arabic, or Chinese. As usual, exponential growth allows the representation of 10^d different numbers with a sequence of only d digits. As for the words of any language, the order of the digits has a precise role in the sequence. Consider the three characters 704 that express the number seven hundred and four. In standard positional notation, each digit contributes a value that depends on its position in the sequence, so 704 indicates seven hundreds, zero tens and four units, and requires that the digit 0 for the tens be explicitly expressed. Furthermore it is understood that the sequence must be read from left to right.

In general people using English or any other European language do not realize how curious the latter request is. If a number is small it can be read immediately from left to right. If however a number is large, for instance 77777777, we must scan it from right to left grouping the digits into threes, as 77,777,777 and only then we are able to read it as seventy seven million, seven hundred and seventy seven thousand, and seven hundred and seventy seven. Not surprisingly, then, most arithmetic operations such as addition, subtraction, and multiplication are also performed from right to left.[2]

It is well known that computers represent numbers in binary, maintaining positional notation, with the digits 0 and 1 accounting for powers of two instead of ten. What is less known is that what was probably the first relevant document on binary numbering is due to the Spanish bishop Johannes Caramuel, who wrote long lists of numbers represented in many different bases.[3]

---

[2]The invention of the zero used today is probably Indian, but was independently used by the Maya in their numerical representations. In medieval times the Europeans learned the importance of the zero from the Arabs and its use rapidly spread. Note that the Arabs write from right to left and numbers used to be read starting from the units, so 704 would sound "four units and seven hundreds." This numerical notation was maintained in Europe where writing went from left to right, possibly because the Roman numbers previously used were written starting with the biggest components at the left. Roman numbers, however, did not raise any problem in reading because different characters were used for units, tens, hundreds etc.

[3]The manuscript *Mathesis Biceps Vetus and Nova* by Johannes Caramuel was published in 1670 and emerged only in 1969 from the archbishopric library of Vigevano in Italy.

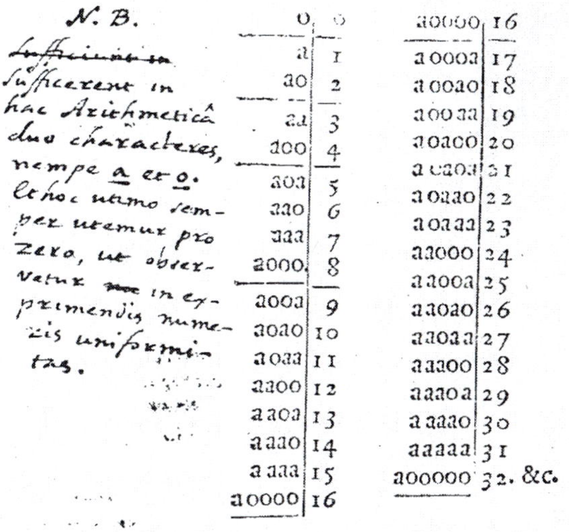

N. B. Sufficerent in hac Arithmetica duo characteres, nempe a et o. Et hoc ultimo semper utemur pro zero, ut observetur in exprimendis numeris uniformitas.

0	0	aoooo	16
a	1	aoooa	17
ao	2	aooao	18
aa	3	aooaa	19
aoo	4	aoaoo	20
aoa	5	aoaoa	21
aao	6	aoaao	22
aaa	7	aoaaa	23
aooo	8	aaooo	24
aooa	9	aaooa	25
aoao	10	aaoao	26
aoaa	11	aaoaa	27
aaoo	12	aaaoo	28
aaoa	13	aaaoa	29
aaao	14	aaaao	30
aaaa	15	aaaaa	31
aoooo	16	aooooo	32. &c.

FIGURE 3.2: A table of binary numbering taken from the *Meditatio Proemialis* in the work *Matesis Biceps Vetus et Nova* by Johannes Caramuel.

Figure 3.2 shows one of his tables of numbers in base two, with the symbol "a" instead of "1."

Computers use a binary alphabet to represent all information according to specific codes. In 1605, Francis Bacon had already discovered that the alphabetic characters could be transformed into sequences of binary digits. More interestingly he noted that the method could be used to represent any object:

" .... provided those objects be capable of a twofold difference only: as by Bells, by Trumpets, by Lights and Torches, by the report of Muskets, and any instruments of like nature."[4]

And in fact it is not only computers that use two symbols to communicate. Two important two-symbol systems were established in the 19th century and survive today. They are Morse code, originally proposed for telegraphy, whose two symbols are *dot* and *line*; and Braille reading for the blind, based on the *presence* or *lack* of a raised dot inside a rectangular region. Another intriguing system is encountered in music, where timelines for percussionists unable to

Numerical bases from 2 to 10, and also larger, are described by Caramuel, together with a passionate philosophical discussion on arithmetic and its relation to music.

[4]Francis Bacon. 1605. *De Augmentis Scientiarum* (On the Proficiency and Advancement of Learning).

FIGURE 3.3: Two timelines in four bars. Bossa Nova as played by Joao Gilberto and Stan Getz, versus Rock as played by The Beatles.

read musical scores are often represented in the so called box notation, with a black square to indicate the time for striking and a white square to indicate a pause. Figure 3.3 shows two well known 4/4 rhythms where the basic timeline is divided into four bars for a total of sixteen equal intervals.[5]

Computer representation of information, Morse code, Braille writing system, and musical box notation, have very different aims and scopes, but share desirable characteristics that seem to be peculiar to all binary systems: coding and decoding of messages is easy, and ambiguities and errors are kept to a minimum. This feature is evident in extreme cases. The Morse code for an SOS message launched by a ship in trouble is a sequence of three dots, followed by three lines, followed by three more dots, where dots and lines correspond to short or long electrical pulses. Such a sequence has a much smaller chance of being misinterpreted than any vocal request for help, particularly in adverse weather conditions.

Similar communication systems have been adopted by generations of inmates in jail, as described by Jack London in a famous novel dedicated to the horror of the prisons of his time.[6] The two symbols used are a knock on the wall or no knocks, as it is practically impossible to rely on their difference in their strength (the situation is similar for Morse code). Various systems have been certainly developed in prison history, possibly without any particular effort to minimize the length of the sequences that, unlike in computer networks, is probably not a priority among inmates.

Computers prefer binary symbols corresponding to the presence or absence of current in a conductor, or to a high or low voltage between two points, independently of their precise values. The reason is as before. A circuit built

---

[5] A comparison of the two rhythms shows the anarchical structure of Bossa Nova (an offspring of Samba) versus the obsessive repetition of Rock. Bossa Nova is strongly syncopated, resulting in one of the hardest rhythms to play. At a first hearing, for example, non-musicians tend to place the second beat at time 5 instead of 4. A top expert of the mathematical properties of musical rhythms is Godfried Toussaint, professor of computer science at McGill University of Montral, Quebec. See the bibliographical notes.

[6] Jack London's *The Star Rover*, written in 1915, is a novel on prison brutality, the death sentence, and reincarnation. A torture device called the "jacket" was used in San Quentin at that time, consisting of a tightly laced canvas garment that compressed the whole body and induced terrible pains.

for deciding between several different events reacts with higher speed and stronger certainty if such events are represented as binary signals rather than by signals that may assume many different values. This happens both in data communication and in data processing, where new signals are generated as the result of a computation on other signals. For example, deciding that some electric current flows in a conductor after the merging of two other currents is much faster than measuring the actual value of the new current. The binary code can be optimized using a minimal number of *binary digits* (*bits*) to represent a number, or a character, or any piece of information, possibly depending on the frequency at which these elements actually appear. We postpone this discussion to section 3.3, turning our attention now to a subtle inconsistency that remains in what we have discussed so far.

All binary systems that we have presented above need an additional feature that may result in the introduction of a hidden third character or some kind of formatting information. For example, consider the sequence "a 0 a 0 a" in Caramuel's binary numbering of Figure 3.2. Without any further indication the sequence is ambiguous because it can represent number 21 or the pair of consecutive numbers 2, 5. The ambiguity could be solved by inserting a *terminator*, say a dollar sign, to distinguish between "a 0 a 0 a $" for 21, and "a 0 $ a 0 a $" for 2, 5. Unfortunately the alphabet has become ternary. This is, in fact, what happens in the Morse code where the transmission includes a "pause" between consecutive characters to indicate termination. For example the aforementioned sequence ● ● ● − − − ● ● ● for SOS is in fact transmitted as ● ● ● $ − − − $ ● ● ● where $ is now the pause, to distinguish it from the sequence ● ● ● − − $ − ● ● ● which means 3B.

In most cases, and in particular in computer data representation, the ambiguity is solved by assigning a field of given length to specific pieces of information. Well known fields are the *byte* (eight bits) to represent a character, and the sequence of three bytes to represent the three components in red, green, and blue of the color of a *pixel* (point on a screen). In the UTF-8 encoding used in the Internet 11100001 is "a," 11000001 is "A," 10110001 is the decimal digit "1," while the group of three bytes 00000000 00110011 11111111 represent a pleasant light blue. Thus formatting allows the unambiguous interpretation of a sequence at the expense of an additional constraint.

As an aside, the Braille system also solves the problem with a formatting rule, as any character is represented in a rectangle of fixed size in which from zero to six raised dots may appear in six specific locations. This gives rise to $2^6 = 64$ combinations to code alphabetic and other typographic characters, plus some special signs to indicate a *change of environment*, for example passing from letters to numbers and vice versa.

We might also ask what nature can teach us in terms of representing information. Until a few decades ago it would have been difficult to give an answer, but in the 1950s biology made a big leap forward with the discovery of the genetic code, although many aspects of this code are still unclear. The famous double helix of DNA molecules, among which other molecules called *bases* line up, code the genetic heritage of all living things, i.e., their *genome* that is in turn divided into one or more chromosomes. A DNA molecule for each chromosome is present in every cell of every organism, with a total of more than three billion bases in a human being. The bases are elementary organic molecules of only four kinds called adenine (A), guanine (G), cytosine (C), and thymine (T).

A standard way of looking at a DNA molecule is considering it as a sequence of bases, abstracting away from the more awkward reality of DNA as a flexible three-dimensional molecule. Since these sequences were fully determined (or the DNA *sequenced*) for many species including humans, a variety of important biological problems have been identified and studied. Each individual has different chromosomes. Chromosomes contain *genes*, i.e., portions of DNA specifying the necessary information to build the proteins that are the main constituents of living bodies. A great majority of genes are the same in a given species, but other parts of the DNA are specific to each individual and allow, for instance, to establish the disputed paternity of a child, or to find the final evidence of a homicide.[7] We will return to this point in a future chapter in connection with networks. For now we concentrate on how the four bases code for proteins.

A simplified view of a protein is as a chain of amino acids that are simple chemical components present in nature in twenty different types. DNA bases code for amino acids, each one of them specified by a triplet of consecutive bases according to a genetic code common to all living organisms and shown in Figure 3.4 (DNA base T is replaced by RNA base U during the biological information transfer). Discovering this code was one of the greatest advances in science of the last century. Clearly the quaternary alphabet of bases allows the formation of up to $4^3 = 64$ triplets, and in fact all of them are "used" in nature with the result that many amino acids have different codes. Furthermore some triplets are signals of the beginning and end of a gene sequence. So for example, the amino acid Serine (shorthand Ser) is represented by six different triplets, while Methionine (Met) is represented by the only triplet AUG that is also the signal of START.

Since the amino acids are always represented by triplets, the code is univocally decodable and, as mentioned before, has a high degree of redundancy whose cause is unknown. Possibly during evolution the number of different amino acids has grown from up to sixteen (two bases required) to the present number, or redundancy reduces interpretation mistakes as in mathematical

---

[7]In February 2010, the 101st child of a *desaparecido* (disappeared) was discovered in Argentina. Thirty-three years earlier his mother had been killed by the brutal military regime immediately after the delivery and the baby was adopted by a "trustworthy" family. The biological father never stopped searching for him together with the "movement of the grandmothers." Thanks to a proof based on DNA similarity he could finally embrace his yet unknown son, crying and laughing at the same time, during a moving TV broadcast. We will talk again of DNA proofs in Chapter 5 as an example of hashing.

BASE 2

	U	C	A	G	
BASE 1					**BASE 3**
U	Phe	Ser	Tyr	Cys	U
	Phe	Ser	Tyr	Cys	C
	Leu	Ser	STOP	STOP	A
	Leu	Ser	STOP	Trp	G
C	Leu	Pro	His	Arg	U
	Leu	Pro	His	Arg	C
	Leu	Pro	Gin	Arg	A
	Leu	Pro	Gin	Arg	G
A	Ile	Thr	Asn	Ser	U
	Ile	Thr	Asn	Ser	C
	Ile	Thr	Lys	Arg	A
	Met	Thr	Lys	Arg	G
G	Val	Ala	Asp	Gly	U
	Val	Ala	Asp	Gly	C
	Val	Ala	Glu	Gly	A
	Val	Ala	Glu	Gly	G

FIGURE 3.4: Table of the genetic code mapping triplets of bases to amino acids.

codes. Sometimes the methods used in nature are too sophisticated to be understood, so DNA remains partly mysterious, and certainly challenging.

Genetic sequences are an important type of data available on the Internet. Operations such as storing and retrieving sequences, comparing sequences for similarities, reconstructing long sequences from fragments, and looking for structural patterns inside a sequence, are essential to modern biology. All these problems share basic principles with others arising in all fields of data processing and related to Web data organization and retrieval.

3.2 Comparing sequences

The body of techniques born for analyzing or giving a form to a sequence of characters, also called *text*, is known as *text-editing*. Typically a short sequence called a *pattern* is searched for in a document. In biological applications the text might be the entire genome of an organism and the pattern a gene whose presence is tested. In a Web page the pattern might be a word and a search engine may be interested in counting how many times it appears. In computer applications sequences of characters are also called *strings* and the above problem is known as *string-matching*.

Formally we want to find all the occurrences of a pattern $P = p_1 p_2 \, p_m$ in a text $T = t_1 t_2 \, t_n$, giving as a result all the positions of T where an occurrence of P starts. From the magic text $T = ABRACADABRA$ and the stupid pattern $P = ABRA$, string matching would yield the result 1, 8, corresponding to the two positions where an occurrence of the pattern starts in the text. Many smart methods have been devised to solve this problem. We start with the most naïve, where a "template" containing the pattern slides along the text, and the positions where text and pattern match are recorded (see Figure 3.5).

$$
\begin{array}{c|ccccccccccc}
 & 1 & 2 & 3 & 4 & 5 & 6 & 7 & 8 & 9 & 10 & 11 \\
T: & A & B & R & A & C & A & D & A & B & R & A \\
P: & A & B & R & A \\
 & & A & B & R & A \\
 & & & \cdot & \cdot & \cdot & \cdot \\
 & & & & & & & & A & B & R & A \\
\end{array}
$$

FIGURE 3.5: The naïve string-matching method.

Observe that all the $n - m + 1$ possible starting points of P in T are considered and, for each one of them, P must be compared with a sub-sequence of T, stopping as soon as a character mismatch occurs. In the example of Figure 3.5 we have $n = 11, m = 4$, and all the positions from 1 to $11 - 4 + 1 = 8$ are considered for a possible start of P. For position 1 a match is found after the comparisons of the four characters A, B, R, A of pattern and text. For position 2 we find a mismatch between $p_1 = A$ and $t_2 = B$, so the comparison is immediately restarted between p_1 and t_3, etc. As we shall see string-matching is a basic problem on the Web.

Many techniques have been proposed to increase the performance of the search, often based on a preprocessing of the pattern in hand as the text is generally not available *a priori*. For example, one can easily observe that finding the pattern $ABRA$ starting in position 1 of the text carries the free additional information that the characters of T in positions 2, 3, and 4 are B, R, and A respectively. Then, as P starts with A, it cannot be present in T starting from positions 2 or 3, and one can restart the search from position 4.

Another important problem in the framework of automatic text editing, and in computational biology, is that of finding the *distance* between two similar but non-identical sequences. Spell checkers, for instance, do not only recognize incorrect words in a document, but also suggest some possible substitutions that are selected among similar words a small *edit distance* from the misspelled one.[8] Formally, given two sequences X, Y we want to find an optimal alignment of X and Y corresponding to a minimal distance between the two. For this purpose blanks may be inserted in each one of the sequences corresponding to the space left by a character wrongly omitted there (*deletion*), or to a character wrongly inserted in the other sequence (*insertion*). Each pair of aligned characters or blanks of X, Y contributes to the distance with a weight 0 in case of match and 1 in case of mismatch, although these costs may be varied at will. A match means that the two characters are equal, keeping in mind that no two blanks are ever inserted in the same positions in X and Y. Mismatch means two different characters, or a character versus a blank. The edit distance is the minimal sum of the weights of all character pairs over all the possible alignments.

For the two alphabetic sequences $X = SUAVE$ and $Y = USAGE$, here are three optimal alignments, all with edit distance three (dashes denote blanks, while plus signs denote mismatches):

S	U	A	V	E		–	S	U	A	V	E		S	U	–	A	V	E
U	S	A	G	E		U	S	–	A	G	E		–	U	S	A	G	E
+	+	+				+		+		+			+		+		+	

Computing the edit distance is not immediate. In order to do it, the two sequences $X = x_1 x_2 \ldots x_n$ and $Y = y_1 y_2 \ldots y_m$ to be compared are formally bordered to the left with a blank (say x_0 and y_0), and their characters are respectively put into correspondence with the rows and columns of an array M of size $(n + 1) \times (m + 1)$. For $X = SUAVE$ and $Y = USAGE$, i.e., $n = 5$ and $m = 5$, we have the array of size 6×6 in Figure 3.6 whose contents is computed as follows.

	–	U	S	A	G	E
–	0	1	2	3	4	5
S	1	1	1	2	3	4
U	2	1	2	3	4	5
A	3	2	2	2	3	4
V	4	3	3	3	3	5
E	5	4	4	4	4	3

FIGURE 3.6: The matrix M of edit distance between prefixes.

Each cell $M[i, j]$ (i.e., the cell in row i and column j) is used to store the edit distance between the two *prefixes* of X and Y ending in positions i and j, that

---

[8]Sometimes with infuriating results.

is, the two subsequences $x_0 x_1 \ldots x_i$ and $y_0 y_1 \ldots y_j$. Row 0 reports the edit distances between the blank prefix x_0 of X and all the prefixes of Y: for example $M[0,3] = 3$ because the edit distance between $x_0 = -$ and $y_0 y_1 y_2 y_3 = -USA$ is three, corresponding to the alignment of the subsequence $-USA$ with a sequence of four blanks (3 mismatching characters). Similarly column 0 reports the edit distances between each prefix of X and the blank prefix y_0 of Y. Formally, we immediately set the values: $M[0,j] = j, M[i,0] = i$, for all i, j.

The value of any other entry $M[i,j]$ can be computed from the adjacent previously computed cells with a smaller value of i and/or j, that is, from $M_1 = M[i, j-1]$, $M_2 = M[i-1, j]$, and $M_3 = M[i-1, j-1]$. Letting $p(i,j) = 0$ if x_i matches with y_j, and $p(i,j) = 1$ otherwise, the elements of M are determined one by one through the following recursive formula:

$$M[i,j] = min\{M_1 + 1, M_2 + 1, M_3 + p(i,j)\} \tag{3.3}$$

where the three options inside the function *min* respectively correspond to increasing the prefix of Y by one position ($j-1$ in $M_1 = M[i, j-1]$ to j in $M[i,j]$) aligned with a blank in X; or increasing the prefix of X by one position aligned with a blank in Y; or increasing both prefixes and comparing the two new characters. The meaning of relation (3.3) should then be clear.

The value $M[n, m]$ in the right bottom corner of the array indicates the edit distance between the two sequences. In the example of Figure 3.6 we have $M[5,5] = 3$ that is the edit distance already found. The optimal alignment (or alignments) that generates the minimal distance can be constructed in a backward procedure, starting from cell $M[n, m]$, tracing back into the array until cell $M[0,0]$ is reached, and deciding at each step the option of relation (3.3) from which the value in the current cell has been derived.

The method for computing the edit distance can be applied with minor modifications to solve a very important variant of the string matching problem already discussed, known as *approximate string matching*. Now the occurrences of a pattern P in a text T are searched for with a certain degree of approximation, for example assuming that such occurrences have a maximum allowed number d of mismatches, insertions, or deletions. In this case T and P correspond to the columns and to the rows of M, but all zeroes are inserted in row 0 to indicate that P can start at any position of T without paying any cost for this. The solution is given in the last row n where the values in each cell $M[n, j]$ express the number of differences between P and the portion of T ending at t_j. There is an approximate occurrence of P in T for each cell of the last row such that $M[n, j] \le d$.

For example the pattern $P = ABCA$ does not occur exactly in $T = ABRACADABRA$, but if we content ourselves with occurrences with at most one error (i.e., $d = 1$) we can build the array M of Figure 3.7 that the reader may examine. In the last row we have $M[4,4] = M[4,6] = M[4,11] = 1$ corresponding to the three alignments with distance 1 of the pattern $ABCA$ with the portions of the text: $ABRA$, $A - CA$, and $ABRA$ again. All these algorithms are essential when searching for portions of text in Web pages, either exactly or approximately.

	–	A	B	R	A	C	A	D	A	B	R	A
–	0	0	0	0	0	0	0	0	0	0	0	0
A	1	0	1	1	0	1	0	1	0	1	1	0
B	2	1	0	1	1	1	1	1	1	0	1	1
C	3	2	1	1	2	1	2	2	2	1	1	2
A	4	3	2	2	1	2	1	2	2	2	2	1

FIGURE 3.7: The matrix M of approximate string matching.

3.3 From sequences to trees

Although all information inside computer memory is represented with a sequence of bits, data with a hierarchical structure are often represented as a *tree*, a very popular structure in computer science. Family trees, like the one shown in Figure 3.8 for the "Buendia," whose epic was narrated in a famous novel of Gabriel Garcia Marquez,[9] are a real-world example of the depiction of data with an inherent hierarchical structure. The need to restrict the representation to certain lines of descendants, the lack of knowledge of some parents, and the multiple marriages of some members of a family may complicate the structure substantially. Computers prefer to deal with cleaner structures. Let us see how trees are encountered in the Internet world.

The terminology is taken from graph theory, botanics, and family trees. The nodes are the elements of the structure corresponding to the names in a family tree, and the arcs are the connections between nodes. The *root* is the node from which all the branches start, José Arcadio Buendia in the example. In computer science trees are considered the other way up to their botanical counterparts, so the root is shown at the top. The *leaves* are the nodes from which no branches start, as for example Amaranta, daughter of José Arcadio and Ursula Iguarán. Parental relationships are the same as for family trees, so one node can be *child, parent, sibling, ancestor*, or *descendant* of another one. A *subtree* is also associated to each node, defined as the portion of tree of which that node is the root.

Although a subsequent chapter will explain this point in more detail, let us

---

[9]Garcia Marquez, G. 1998. One Hundred Years of Solitude, Harper Perennial, New York. Figure 3.8 is an elaboration of the Buendia family taken from Wikipedia at: http://en.wikipedia.org/wiki/One_Hundred_Years_of_Solitude

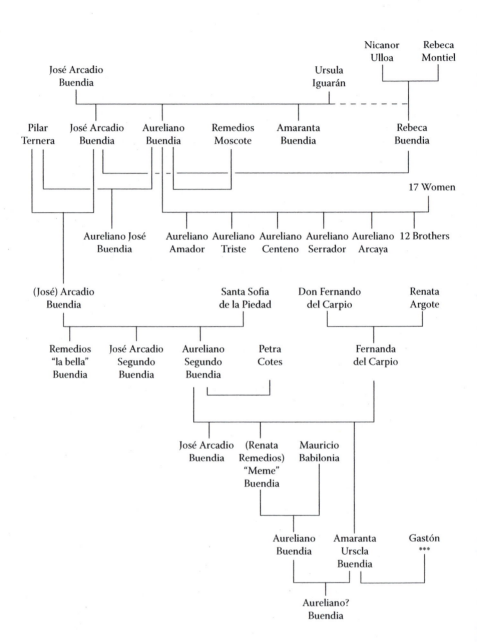

FIGURE 3.8: The Buendia family tree.

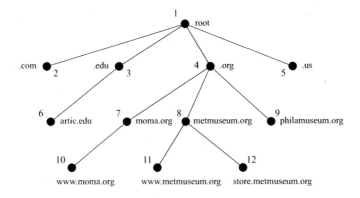

FIGURE 3.9: A (very small) portion of the *DNS* tree with the *names* of some American museums.

consider something that is probably known to most readers, at least something which they have exploited whilst using the Internet. In order to be reached by their peers, all the devices of a network must have an address, called an *IP* (for *Internet Protocol*) *address* on the Internet.[10] Devices, or the information contained in them, also have names, called *domain names* on the Internet. Essentially a domain name indicates what we are looking for and the corresponding IP address indicates where it is. An efficient tool must be used to map names to addresses and to retrieve an address from the corresponding name. This tool is the *DNS* (for *Domain Name System*) *tree*, a huge data structure that is replicated in many computers called *name servers*. Any device looking for a resource in the network only needs to know the name of the resource and the address of a Name Server.

Being organized in the *DNS* tree the names have a hierarchical structure that makes it possible to access them very fast. The node at the root, labeled with a silent dot, is the most important of all as we shall see in a later chapter. The children of the root are named .com, .edu, .org etc. to denote particular categories of users, or are named .us, .ca, .uk, and recently .eu for Europe etc. to denote different countries. Going down the tree, each successive level inherits the name of the parent with an additional word to left of the label, with all these words separated by a dot. For example www.metmuseum.org is the name of the Web server of the Metropolitan Museum in New York, where .org denotes a non commercial entity and metmuseum.org is the museum name. Part of the *DNS* tree is shown in Figure 3.9.

Several methods exist to represent a tree in the form of a sequence. Purists may prefer a parenthesized sequence where the tree is recursively coded as the (name of) the root, followed by a pair of open and closed parentheses that

---

[10]Presently an *IP* address is a 32 bit number.

enclose the children of the root, each recursively represented with the same method. For example the sequence:

$$1 (2 \; 3 (6) \; 4 (7 (10) \; 8 (11 \; 12) \; 9) \; 5) \qquad (3.4)$$

corresponds to the tree of Figure 3.9 with nodes represented as integers for clarity. The root 1 is followed by the list in parentheses of its children 2, 3, 4, 5, all immediately followed by the list of their children. For example node 3 is followed by the list of its only child 6; node 4 is followed by the list of its children 7, 8, 9; and so on. The reader may verify that at most $2(n-1)$ parentheses are necessary to represent any tree of n nodes.

Depriving sequence (3.4) of its parentheses we obtain the new sequence:

$$1 \; 2 \; 3 \; 6 \; 4 \; 7 \; 10 \; 8 \; 11 \; 12 \; 9 \; 5 \qquad (3.5)$$

which is insufficient to reconstruct the tree univocally. However, this sequence is still strongly related to the tree, because it gives the order in which nodes are examined in what is called a *pre-order traversal*. Intuitively the rule is the following. Assume that the tree is *ordered*, that is, a left-to-right order is always defined among siblings and is reflected in the graphical representation of the tree. Starting from the root go down the tree as far as possible, always taking the left branch and leaving open any alternative at nodes that have more than one child. When a leaf is reached, backtrack to the last open point and follow the next alternative available. For a node N with $h > 0$ children $C_1, ..., C_h$ the rule is *recursive*, as shown in Figure 3.10. For the non expert the rule may need a little consideration to understand how recursion works.

PRE-TRAV(N):

examine(N);

PRE-TRAV(C_1); ....; PRE-TRAV(C_h).

FIGURE 3.10: The recursive rule for pre-order tree traversal, triggered by $N = $ root.

It has often been observed that the pre-order traversal induces on the nodes the same order of succession to the throne that applies to the members of a royal family.[11] Take Figure 3.9 as the representation of a royal family tree with node 1 as the king. Assume the whole family gathers in a castle where each current king is assassinated every hour on the hour. The crown passes from the late king 1 to prince 2, then to his sibling 3 as 2 has no descendants, then to 6, 4, 7 etc., until the whole progeny is reset to zero in twelve hours.

Besides representing a hierarchical data organization, trees are widely employed in computer science to model different objects or to depict different situations. After all a tree can be seen as a simple type of connected graph with n nodes and $n-1$ arcs. This many arcs are necessary and sufficient

---

[11]e.g., see D. E. Knuth. *The Art of Computer Programming*, Vol. 1 p. 335.

to keep the tree structure connected. To see this, observe that for $n = 1$ or $n = 2$, zero or one arc is required. Inductively if $n - 2$ arcs are necessary and sufficient for connecting $n - 1$ nodes, connecting the n-th node to the subgraph through a minimum number of arcs requires one and only one new arc, thereby increasing the number of nodes and arcs to n and $n - 1$, respectively. Therefore computer networks can be organized in the form of a tree if it is important to keep a low number of connections, although this structure is very fragile because the failure of a single arc is sufficient to disconnect the network.

Trees also play an essential role in the creation and handling of electronic documents to be published on the Web, including image or audio-video files. For this purpose a *markup* language called XML is becoming the standard. An XML document is a sequence of characters of two different types, namely, the ones specifying the content of the document and the markup characters.

A markup sub-sequence starts with "<" or "&" and stops with ">" or ";" respectively. Sub-sequences that are not markup are of content type. Consider the portion of XML program in Figure 3.11. The words "artgallery," "painting," "caption," and "date" included in markup sub-sequences are called *tags* and are self-explaining. A bar as in "/artgallery" indicate the end of a tag scope. The word "img" is also a tag defining an image, and contains the two attributes, "src" and "alt" specifying the source file of the image and an alternative text in case the image is not available (in this case the tag ends with "/>").

```
<artgallery>
    <painting>
        <img src="Painting1.jpg"
            alt="Panic Room, by Peter Halley"/>
        <caption>Peter Halley's "Panic Room", painted in
            <date>2005</date>.
        </caption>
    </painting>
    <painting>
        <img src="Painting2.jpg"
            alt="Indexed, by Peter Halley"/>
        <caption>Peter Halley's "Indexed", painted in
            <date>2007</date>.
        </caption>
    </painting>
</artgallery>
```

FIGURE 3.11: Portion of an XML program depicting the contents of an art gallery.

It is not difficult to see that our XML program is none other than a tree, where the first tag represents the root and the beginning of a new tag corre-

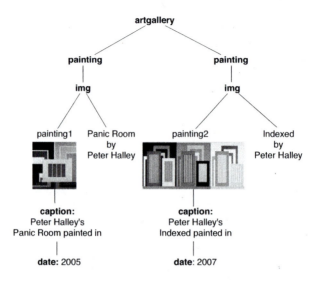

FIGURE 3.12: The binary tree induced by the XML program of Figure 3.11. Tags are in boldface. The two images, painting1 and painting2, are reproduced with the kind permission of Peter Halley.

sponds to the insertion of a new subtree. The repetition of a previously used tag specifies the insertion of a new subtree at the same level, that is, a subtree rooted at a new sibling. The tree corresponding to the given example is shown in Figure 3.12.

In addition to trees in the traditional sense of the word, many other tree-structures are found in nature, in particular when a process of growth develops by dichotomy. Corresponding to this case, computer scientists have introduced a more constrained structure called a *binary tree*, where each node has at most two children and each of them has the specific role of *left* or *right* child. So a node may have only one child that may be a left child or a right child, and these are treated as two different conditions. Binary trees are of paramount importance in search problems, as we shall see in the next chapter. A different application has to do with text coding and compression, and is crucial for the functioning of Web search engines. To discuss it we must restart from the problem of the unambiguous interpretation of sequences that we have solved thus far only through formatting or insertion of separators.

In a binary representation of the English alphabet where A, B, and C are respectively coded with 0, 1, and 01, the sequence 0101 admits the four interpretations: ABAB, ABC, CAB, CC. If the space is also coded with 10 the sequence is still ambiguous and admits the further interpretation A_B. Many unambiguous binary codes have been invented, among which *prefix codes* are of particular interest. In these codes the sequence of bits associated to

Char	space	t	i	a	n	s
Freq	4	4	3	3	3	2
Code	000	001	100	101	110	0100

Char	o	c	'	g	f	e
Freq	2	2	1	1	1	1
Code	0101	111	01100	01101	01110	01111

FIGURE 3.13: Character weights (Freq), and their codings (Code) generated by the Huffman tree.

each symbol is never a prefix of the sequence of any other symbol, so that unambiguous decoding can be performed on the fly while scanning the text from left to right. For example, the coding A=0, B=10, C=11 is prefix free and the sequence 10011010 admits the only interpretation, BACAB, that is directly found during the scanning. The coding A=0, B=01, C=11 is also unambiguous but not prefix free (A is a prefix of B) and the sequence 011111 can be univocally interpreted as BCC only after having scanned it completely (deleting a 1, the sequence would be ACC).

A fundamental contribution to coding theory was the prefix code proposed by David Huffman in 1952, still widely used today as a basic tool for text compression. The coding of each character of a text is represented with a number of bits that is a function of the probability of the character occurring. The higher the probability, the fewer bits are used. In this way the total length of the coded text is minimized over all possible character codes.[12] Huffman gave a very efficient coding method which attains on average the minimum length of the coded text if the probability of each character occurring is known a priori, and can be adjusted adaptively on observed frequencies if the probabilities are unknown.

Each character has an associated *weight* equal to its probability of occurring or to its frequency in the text if this is known. For instance the Rolling Stones' sentence "I can't get no satisfaction" consists of twenty-seven characters that appear with different frequencies. The most frequent are *space* and *t* with four occurrences, followed by *a*, *i*, and *n* with three, and so on. These frequencies are taken as weights and the characters are sorted in non-increasing order of the weights as shown in Figure 3.13 (rows "Char" and "Freq"). Let us follow the Huffman tree construction in Figure 3.14. First, the characters are associated to the leaves of the tree to be built. Then the two

---

[12]Note the similarity with natural languages where common words tend to be very short. In fact the concept of character is taken in a broad sense in the coding and can refer to a whole word in an "alphabet of words."

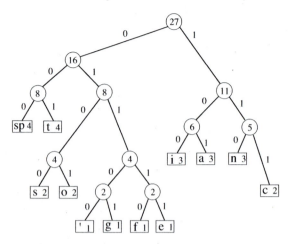

FIGURE 3.14: The Huffman tree for the sentence "I can't get no satisfaction."

leaves X, Y with smallest weights are considered and a new internal node Z is created as their parent. If X and Y have different weights, the one with greater weight is taken as the left child of Z. The sum of the weights of X, Y is assigned to Z as its weight, and bits 0 and 1 are assigned to the left and to the right child of Z, respectively. In our example, the leaves $X = {}'$ and $Y = g$ with weight 1 are selected, a new internal node with weight 2 is created as their parent, and the bits 0 and 1 are put on the arcs leading to ${}'$ and g. Now another internal node is created, having as its children two nodes with the smallest weights among the remaining leaves and the already created internal node, with the child of larger weight on the left. In the example, the new node is the parent of f and e (weight 1) and has weight 2. Then s and o both with weight 2 are selected to form a new node of weight 4. Then, the two nodes selected are the two internal nodes with smallest weight 2. The procedure is iterated until all existing nodes have been connected in pairs and only one node remains at the root. The coding of each character x is then generated following a path from the root to the leaf containing x and assigning 0 for each step down to the left, and 1 for each step down to the right. In our example the resulting coding sequences are shown in the row "Code" of Figure 3.13. It is easy to see that this procedure gives rise to a prefix code.

Note that the Huffman tree is a particular binary tree with all internal nodes having exactly two children. This implies that for h internal nodes we must have $h + 1$ leaves, as one can easily prove by induction (in our example we have ten internal nodes and eleven leaves). So the *total* number of nodes

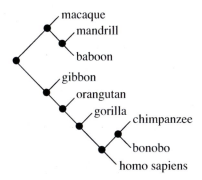

FIGURE 3.15: A phylogenetic tree for a set of *catarrhine* primates. *Homo sapiens* (literally, knowing man) are the humans.

is $n = 2h + 1$. With some ingenuity it can be verified that the time required by the above procedure is proportional to n.[13]

We conclude this roundup of trees as we did for sequences, by looking at a further connection between computational science and biology. This will also show an unexpected linkage between edit distance and prefix code construction, a further indication of the general validity of these concepts.

An important tool for studying the evolution of different biological species is the *phylogenetic tree*, a binary tree containing existing species at the leaves and pairing nodes to form inferred (i.e., not direclty observable) internal nodes as probable most recent common ancestors. An example of such a tree for some Old World monkeys is shown in Figure 3.15. The internal nodes (black dots) are unknown ancestors from which the different species differentiated in time, with the most ancient at the root.

Phylogenetic trees have been built for some time on the basis of morphologic or genetic characters, and more recently by the application of DNA sequences. We recall only some basic points, leaving any further investigation to the bibliographic notes. One can make use of a *character matrix* that reports the "values" of a set of characters for any of the species under consideration, as in the toy example of Figure 3.16 where five species are indicated with the Greek letters α to ϵ, and their five characters c_1 to c_5 have DNA bases as values. Assuming that the value of a character changes to differentiate two species, a branching denotes one such event, the corresponding internal node is labeled with the evolving character name, and the two branches carry the two values of the character that (hopefully) will apply to all the descendants of that branch. If this is true, that is if that character will not change again in evolution, the resulting tree corresponds to a *perfect phylogeny*. The character

---

[13]Such considerations will be developed in the next chapter. It should be clear, however, that a procedure that builds a tree of n nodes in time proportional to n is *optimal* in its order of magnitude, as such a time cannot be beaten by any other procedure.

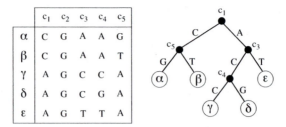

	c_1	c_2	c_3	c_4	c_5
α	C	G	A	A	G
β	C	G	A	A	T
γ	A	G	C	C	A
δ	A	G	C	G	A
ϵ	A	G	T	T	A

FIGURE 3.16: An example of perfect phylogeny. A matrix for species α to ϵ and characters c_1 to c_5; and a corresponding phylogenetic tree.

matrix of Figure 3.16 is in fact consistent with a perfect phylogeny, corresponding to the tree shown in the figure. When a branching is established, as for example at the tree root for character c_1, all the nodes of the left (respectively, right) subtree have $c_1 = C$ (respectively, $c_1 = A$), as verified for the actual species at the leaves.

Unfortunately the character matrix of most observed species is not consistent with any perfect phylogeny. In our example it is sufficient to change one nucleotide to make a perfect phylogeny impossible. For instance, replace ϵ with a new species ζ with character values A G T A A (i.e., change the forth nucleotide). As before character c_1 would induce a partition between α, β on one side, and γ, δ, ζ on the other side. But character c_4 would also induce a partition between α, β, ζ on one side, and γ, δ on the other side, creating an inconsistency for ζ. As in many other cases, particularly when dealing with life sciences, a compromise must be reached.

To this end the most common approach is based on a *distance matrix* that reports the "distance" between any two species under a chosen metric. For instance taking a DNA subsequence as a vector of character values for each species, the distance between two species can be computed as the edit distance between such subsequences as defined in section 3.2. For the example of Figure 3.16 with species ζ instead of ϵ we would have the distance matrix of Figure 3.17: e.g., the best alignment between α and ζ is given by: C G - A A G versus A G T A A -, with edit distance 3. The most common method for building a phylogenetic tree based on a distance matrix is known as UPGMA (for *Unweighted Pair Group Method with Arithmetic Mean*). The reader may notice a strong similarity with the Huffman tree construction.

Starting with the species at the leaves, an internal node is formed to connect the two species with minimum distance, α and β in the example. A new entry called *cluster* is created to substitute the paired nodes in the matrix, and is associated with the new internal node. Then the matrix values for the cluster are recomputed, and new clusters are similarly built one after the other until the matrix collapses to a unique entry. In this process the distance between two clusters is simply computed as the average of the distances of

α	β	γ	δ
β	1		
γ	4	4	
δ	4	4	1
ζ	3	3	2	2

αβ	γ	δ
γ	4	
δ	4	1
ζ	3	2	2

αβ	γδ
γδ	4
ζ	3	2

FIGURE 3.17: The distance matrix for the example of Figure 3.16 with species ζ instead of ϵ, reporting the edit distances between the sequences of character values. Successive transformations of the matrix with emerging clusters, according to the UPGMA method.

all the pairs of elements in the first and the second cluster. In our example cluster $\alpha\beta$ is constructed first, and its distances with the other elements of the matrix are evaluated as: $\text{dist}(\alpha\beta, \gamma) = \text{dist}((\alpha, \gamma) + \text{dist}(\beta, \gamma))/2 = 4$, and similarly $\text{dist}(\alpha\beta, \delta) = 4$, and $\text{dist}(\alpha\beta, \zeta) = 3$, as reported in the second instance of the matrix. Then leaves γ and δ are paired in the new cluster $\gamma\delta$ and the distances of this cluster with $\alpha\beta$ and ζ are recomputed in the third instance of the matrix. Then $\gamma\delta$ is paired with ζ, and a final root will connect the two clusters $\alpha\beta$ with $\gamma\delta\zeta$. It may be noted that, although a perfect phylogeny is not possible now, the resulting tree is still that of Figure 3.16 with ζ in the place of ϵ.

Bibliographic notes

The first work to be mentioned here is: Knuth, D.E. Started 1968. *The Art of Computer Programming*. Addison-Wesley. The complete work is planned to be in seven volumes, of which Vol. 1, 2, 3 have been available for a long time, Vol. 4 is partly available in fascicles, and Vol. 5, 6, 7 are planned. Possibly this is the most fundamental book in computer science, excellent both for study and for consultation. Any computer scientist should keep a copy of it in their personal library. In particular Vol. 1 and in part Vol. 3 gives most of the mathematical basis for the present chapter. Further reading could include any of several books on algorithms that we will comment on in detail in chapter 4.

For studying matching problems between combinatorial structures, an essential volume is Gusfield, D. 1997. *Algorithms on Strings, Trees, and Sequences*. Cambridge University Press. Concepts of coding and information theory can be studied in depth in the excellent book: Cover, T.M. and J.A. Thomas. 2006. *Elements of Information Theory*. Wiley-Interscience; or in other texts available on line.

The relations between genetic information and computational science can be studied, for example, in one of the first books on Bioinformatics: Setubal, J. and J. Meidanis, 1997. *Introduction to Computational Molecular Biology.* PWS Publishing Co.

The reader interested in the mathematical properties of musical rhythms can start, for example, looking at the presentation of G. Toussaint in the proceedings of the biennial conference: *Fun with Algorithms*, Edition Plus, Pisa, Italy 2004. A very good view from a musical perspective can be found in: Klver, T. 1977. *The Joy of Drumming: Drums and Percussion Instruments from around the World*, Binkey Kok Publishing, Diever, Holland.

Chapter 4

The algorithm: the key concept

How shameless Martin applies mathematical logic when courting girls, and how a police officer may be unable to identify a criminal gang in the Web.

The algorithm[1] is the key concept of this book. When in the 1930s Alan Turing and other logicians came up with a sound definition of the algorithm, computer science took its first steps, although no computers had yet been invented. Algorithms accompanied the growth of this new science up to the Internet and through to the Web era where they lie at the foundation of everything. Based on the primitive concepts discussed in the previous chapters we will now illustrate several algorithms to be used later as building blocks, starting from the indispensable theoretical foundations of our discussion.

Let us return to the world of sequences. Our analysis will not lose generality if we restrict this world to the binary case that interests us most. Call these sequences $\sigma_0, \sigma_1, \sigma_2, ...$ and arrange them in *canonical order*, i.e., by increasing length and, for the same length, by increasing value of the corresponding binary numbers, starting from the empty sequence ϵ. We have:

$$\begin{array}{cccccccccccc} \sigma_0 & \sigma_1 & \sigma_2 & \sigma_3 & \sigma_4 & \sigma_5 & \sigma_6 & \sigma_7 & \sigma_8 & ... & \sigma_{14} & \sigma_{15} & ... \\ \epsilon & 0 & 1 & 00 & 01 & 10 & 11 & 000 & 001 & ... & 111 & 0000 & ... \end{array} \quad (4.1)$$

i.e., the binary sequences can be put in a one-to-one correspondence with the natural numbers 0, 1, 2, ..., or, in the terminology of set theory, such sequences are *numerable*. Adopting the numbering induced by the canonical order, for any given sequence σ_i we can reconstruct in finite time its number i, and vice versa, no matter how long the sequence is. Keep in mind that we are talking of sequences of *finite* but *unbounded* length, that is, for any length l arbitrarily fixed there are (infinite) sequences longer than l.

This infinite world of numerable objects has attracted thinkers since ancient times. The mere existence of infinity was a matter of discussion until

---

[1] A book on the algorithmic foundations of networking must include a chapter like this one. Readers with some knowledge of computability and complexity theory, as acquired, for example, in a standard university course in computer science, may skip the chapter or just skim it to refresh the memory.

Aristotle asserted beyond any (personal) doubt that infinity had to exist: for example, he said, numbers never finish. In the western world Galileo Galilei was the first to take up this concept again in his last and perhaps most important work entitled "Dialogues and Mathematical Demonstrations Concerning Two New Sciences" (the book was published in 1638 in Leiden out of the reach of the Inquisition, while Galileo was serving a life sentence for heresy under house arrest near Florence). In a deep speculation on viewing a geometric shape as an infinite collection of points of infinitesimal size, Galileo pointed out some unexpected facts that arise when dealing with infinity, suggesting in particular the "paradox of squares."[2] Each positive integer, he noticed, has a square obtained by multiplying the number by itself. So there are as many integers as squares. However there are an infinite quantity of integers like 3, 5 etc. that are not squares of anything, so one should also conclude that there are more integers than squares. He explains that "these difficulties arise when discussing the infinite with our finite minds," while the facts observed are perfectly normal. And in fact all sort of paradoxes of this type can be conceived.

Consider for example all the positive rational numbers, i.e., the ones expressed by the ratio of two positive integers, plus the zero. A canonical order can also be defined here by arranging the rational numbers τ_i by increasing value of the sum of the numerator and the denominator, and, for the same value of this sum, by increasing value of the numerator. We have:

$$
\begin{array}{ccccccccccccc}
\tau_0 & \tau_1 & \tau_2 & \tau_3 & \tau_4 & \tau_5 & \tau_6 & \tau_7 & \tau_8 & \tau_9 & \tau_{10} & \cdots \\
0 & 1/1 & 1/2 & 2/1 & 1/3 & 2/2 & 3/1 & 1/4 & 2/3 & 3/2 & 4/1 & \cdots \quad (4.2)
\end{array}
$$

so also the rational numbers are numerable. That is, they are as numerous as the natural numbers. Note that each element in the second line of (4.2) has infinite repetitions of its value, for example the value of $1/1$ appears also as $2/2$, $3/3$, ...; the value of $1/2$ appears also as $2/4$, $3/6$,..., etc. Nonetheless, even if all these repetitions are taken away and the second line is compacted by shifting its elements to the left to fill all the empty positions, we end up with two lines with the *same number* of elements, but where the second line contains all the values of the first line ($1/1 = 1$, $2/1 = 2$, etc.) plus many others ($1/2$, $2/1$, $1/3$, etc.). A paradox, then!

Although Galileo had already explained that these apparent contradictions are due to the incorrect application of ordinary reasoning to infinite objects, it was not until the late 19th century that a sound theory of infinite sets was given by Georg Cantor. This theory is completely abstract but has produced some practical consequences of paramount importance in the world of algorithms. First Cantor proved that there is no inconsistency in having two

---

[2]The word paradox, etymologically "against the (current) opinion," should strictly speaking only be used in this sense, as in the paradox of squares that describes a true but counter-intuitive situation (there are as many integers as squares). In mathematics the meaning has more recently been extended to entail antinomy, etymologically "against the rule," i.e., a statement contradictory in itself, e.g., for containing two mutually exclusive assertions.

infinite sets A, B of the same *cardinality* where A is a proper subset of B, as for the integers and the rationals of (4.2). The question then arose whether all infinite sets have the same cardinality, and the somewhat surprising answer was negative. In particular, Cantor proved that the real numbers are more numerous than the integers, that is, that there is no way of numbering the reals. Using the same reasoning one also can prove that mathematical functions are *too many* to be numerable, as we will explain below. The poorest, least complicated infinity is that of the integers where computer science is fully immersed.

4.1 Functions, algorithms, and decidability

The language of mathematics, like any other reasonable language, is aimed at describing facts using sequences of characters taken from a finite alphabet. All known mathematics, from Apollonius' conics to Bourbaki's set theory; all lost mathematics, from Euclid's surface loci to Riemann's notes; and all not yet born mathematics that one can think of; can be recorded in one very large book according to some typographical rules. This language, then, follows the general rules explained in Chapter 3 for the sequences, whose number grows exponentially with their length. And since such a length is a priori unbounded, the number of sequences with a mathematical meaning is potentially infinite. Still, being sequences they are at most as many as the natural numbers, i.e., they are numerable.

Consider now a mathematical function $y = f(x)$. As we know this expresses a correspondence between two possibly infinite sets, a *domain* X and a *co-domain* Y. In the simplest case X and Y are numerable, then all the elements $x \in X$ and $y \in Y$ can be represented with finite sequences giving their names or values. In the language of computers, the computation of f on the datum x produces the result y. If an algebraic or any other unambiguous expression for f exists, as for example "$y = 2x^3 + 1$" where x and y are integers, or x and y are English words and "y is obtained from x by replacing all a's with b's," then f itself can be expressed with a finite sequence although its domain and co-domain are infinite. But this is not always the case. Cantor's theory implies that the functions are not numerable, then there are not enough finite statements (expressions, rules, names, or whatever else) to represent all possible functions. In the absence of a finite expression, the only way of representing a function would be to give two lines that explicitly show the elements of X and the corresponding elements of Y. However these lines would have infinite length, and Cantor's theory shows that all the infinite sequences of elements taken from an infinite set, Y in this case, are non numerable: this is why the functions are non numerable. The crucial reason why we are saying all this will be now made clear.

Informally an algorithm is an unambiguous procedure to solve a specific problem on arbitrary data using the rules of the game at hand: for example, producing the pre-order list of the nodes of an arbitrary tree using a given computer, according to the recursive rule given in Chapter 3; or making mayonnaise with an arbitrary number of ostrich eggs using a given blender, according to the recipe of a sub-Saharian cookbook.[3] After the very first steps of the computing era, i.e., starting in the early 1960s, the recognition of the importance of designing algorithms independently from the machine then used to execute them gave rise to the birth and subsequent growth of the huge field of programming.

Today computer algorithms are formulated in any reasonable language that captures their main features of interest, and then implemented in programming languages accepted by all computers. No matter how it is expressed, an algorithm transforms a datum x (the tree or the eggs) into a result y (the pre-order list or the sauce), so it has the same role and power as the computation of a function. And here a crucial consequence of Cantor's theory becomes pertinent. Like many other mathematical objects, algorithms are sequences of finite length, and hence are numerable. But functions are not numerable, hence there must be *non computable functions*, or equivalently, well-posed problems that cannot be solved algorithmically. And since there is nothing inherently against the fact that a non computable function may be defined with a finite expression, logicians started hunting for such functions just after Cantor's theory became accepted. The first consequence, however, was the birth of a rigorous definition of algorithm that did not exist before.

Stepping back a little, it is not surprising that a formal study of algorithms came out of a search for a negative result, i.e., the search for a function not amenable to algorithmic computation. Whenever a problem could be mathematically solved, like computing the square root of a number or constructing the bisector of an angle, the algorithm was simply described without particular attention to the language used and, in most cases, even to the efficiency of the suggested procedure. This is what we have done in Chapter 3 when explaining how to determine the edit distance between two sequences in mathematical terms, or which rule to follow for traversing a tree. However, when it comes to stating that a particular problem *cannot* be solved algorithmically, a precise definition of algorithm must be given in order to prove that no algorithm can help.

Many logicians worked on this in the first decades of the 20th century, and all of them should be given due credit for the development of the field. In 1936, the British mathematician Alan Turing gave a historic account of an abstract machine with an infinite tape memory, now universally known as a Turing Machine or TM, together with a problem that cannot be solved by such a machine. The TM is now accepted as a basic computational model, and any known computer algorithm can in fact be reformulated as a TM. The famous

---

[3]Several procedures described informally in the previous chapters are indeed algorithms.

halting problem proved by Turing to be algorithmically unsolvable has had a wealth of consequences in the programming world. The reader interested in knowing more of this theory is referred to the bibliographical notes. We merely recall here a few details to cast light on the great lesson Turing left us.

Since a TM is an algorithm we will speak equivalently of algorithms or TMs, and of data or sequences written on the machine tape. Being an algorithm, a TM is *less* than a computer, but one can define a *universal* TM U able to simulate the computation of any other TM T on its data D, provided that T and D are given to U as input. In a sense U is the equivalent of a computer. The simulation is possible because algorithms (or TM descriptions) and data are formulated with the same alphabet, so the sequence that specifies T can be treated as a datum by another TM, or even by T itself. The latter observation brings us into the treacherous world of *self reference*, a beloved way of reasoning for any logician and a key ingredient for proving the impossibility of drawing consequences from a sound basis.

The story started long ago with the epistle of Saint Paul the Apostle to Titus, whom he had left in Crete to set things in order. Unfortunately the holy man was not known for being particularly kind or politically correct. So, after having alerted Titus that in Crete "there are many insubordinate, both idle talkers and deceivers, especially those of the circumcision," he added (Ch.1 - 12):

One of them, a prophet of their own, said, "Cretans are always liars, evil beasts, lazy gluttons."

The prophet is identified as Epimenides, a mythical Cretan sage to whom Saint Paul attributes the self-referential statement of being a liar. The statement, later known as the *liar's paradox*, is also self contradictory.[4] Coming from a Cretan liar the statement should be false, thus implying that the Cretans do not lie. In this case, however, the statement should be true thus implying that the Cretans lie. In mathematical terms the statement is true if and only if it is false: an antinomy. Not only the Cretans, but the laws of thought seem to have been ill-treated. The search for non-computable functions uses the same logical structure, that is, the assumption of a certain problem being algorithmically solvable leads to an antinomy. We sketch here the first proof given by Turing. Readers with more practical interests may skip the proof and go directly to its consequences.

We shall restrict our discussion to decision problems that require a yes/no answer (or to functions with a binary co-domain), noting that they retain the inner nature of the whole class of problems amenable to algorithmic solutions. In most applications these problems are concerned with deciding whether a solution with certain characteristics exists or not rather than giving the solution itself, as for example for the Hamiltonian tour in a graph discussed in Chapter 2, where deciding if such a tour exists, or actually finding it,

---

[4]See footnote 2 for the meaning of the word paradox.

requires essentially the same computations. In general a *decision algorithm* A for solving one such a problem produces an affirmative result **yes** for some input D' (a graph having a Hamiltonian tour), but a negative result **no** for some other input D'' (a graph without such a tour), but may also run for an infinite amount of time without ever producing a result. This may be caused by a trivial formulation error that leads the computation into an endless cycle; but may be subtly unavoidable, for example, when testing a possibly infinite set of sequences to find out whether there is one with a certain property.

Denoting with $A(D)$ the computation of A on D, and with $A[D]$ its result, we have $A[D] = $ **yes**, or $A[D] = $ **no**, or $A(D)$ does not terminate. Rephrasing Turing's reasoning we then pose the following:

Halting Problem. *Given an arbitrary algorithm A and an arbitrary datum D, decide in finite time whether $A(D)$ terminates.*

Although very simple and logically legitimate, the halting problem is *undecidable* in the sense that there is no algorithm T for deciding in finite time if $A(D)$ terminates for any pair A, D taken as input. The proof is by contradiction. Let such a T exist. Take two arbitrary sequences S, Q. The computation $S(Q)$ has a meaning if S is a legitimate coding of an algorithm (we can say, if S is an algorithm). Then we can define the behavior of T as:

$$T[S, Q] = \textbf{yes} \text{ if } S \text{ is an algorithm}$$
$$\text{and } S(Q) \text{ terminates;} \tag{4.3.i}$$

$$T[S, Q] = \textbf{no} \text{ if } S \text{ is not an algorithm}$$
$$\text{or } S(Q) \text{ does not terminate.} \tag{4.3.ii}$$

Note that T must always terminate in finite time; hence it cannot simply consist of the simulation of S on Q because such a simulation may not terminate. In a sense T is a monitor of the behavior of its peers. Consider now the specific algorithm ANT (for antinomy) with arbitrary data D:

$ANT(D)$
    if $(T(D, D) = $ **yes**) **loop forever**
    else answer yes;

$ANT(D)$ does not terminate if and only if $T(D, D) = $ **yes**; i.e., by relations (4.3) D is an algorithm and we can write: $ANT(D)$ does not terminate if and only if $D(D)$ terminates. But since ANT is an algorithm we can execute it on $D = ANT$ thus obtaining:

$ANT(ANT)$ does not terminate if and only $ANT(ANT)$ terminates. (4.4)

So, we have built an antinomy. As the only unproved step in our logical chain of deductions is the assumption that T exists we must conclude that this is actually impossible. With the same reasoning a wealth of other undecidable problems have been found, such as the so called *Universal Problem* of deciding if $A(D) = $ **yes** and the like. The lesson of Turing is the impossibility of deciding, among peers, the results that will be obtained by others, even if we know everything about them (T knows A and D). Deciding whether a TM will eventually halt requires a more powerful machine than another TM. Once a peer is known one can simulate his behavior, as the universal TM does; but while the simulation is ongoing it is impossible to foresee if it will ever stop.

In practice it is impossible to write a general program for deciding whether other programs are written correctly, for example to see if they will get stuck in an endless loop for some input data. The science of programming has made impressive advances in checking program correctness, but absolute guarantees are theoretically impossible. And now let us see how a shameless womanizer named Martin has fully learned the lesson.

Martin is able to do something I am incapable of. Stop any woman on any street.

This is the *incipit* of a famous novel. To proceed, Martin applies to the women a technique coincident with the one usually applied to formal systems in mathematical logic.[5] He is at the same time a datum for countless algorithms, and an algorithm for countless data. Fundamentally the situation is simpler than it may appear. Martin approaches all women with the aim of seduction and the women, now algorithms, may accept, or refuse, or postpone a decision thus leaving him in complete uncertainty. But Martin is too smart to wait forever; so he goes on developing parallel courting actions in different stages of advancement.

In the first stage, that Martin calls sighting, he simply records the names of the women that he may be seducing one day. So this is essentially an enumeration procedure (although apparently countless, women are clearly numerable). According to him this stage is of maximum importance because:

From his vast experience, he has come to the conclusion that it is not as difficult, for someone with high numerical requirements, to seduce a girl as it is to know enough girls who he hasn't yet seduced.

Then he lets all adventures to proceed in parallel independently from one another, until some of them will come to a positive conclusion without any waste of time. The technique is based on the *diagonal trajectories* used by Cantor in his theory of infinite sets, although we believe that Martin has reinvented them by himself.

Martin's and the girls' behavior is illustrated in the table of Figure 4.1. Girls have been already registered (and numbered) in a stage zero not shown. **Yes** and **no** are final girls' decisions that halt their algorithms, and **wait**

---

[5]Martin is the main character of "The golden apple of eternal desire," in the collection of stories *Laughable Loves* by the Czech writer Milan Kundera, 1968. We refer to the English translation of Suzanne Rappaport, published in the USA by Alfred A. Knopf, Inc., 1974. Kundera seems to be unaware of the mathematical capabilities of Martin.

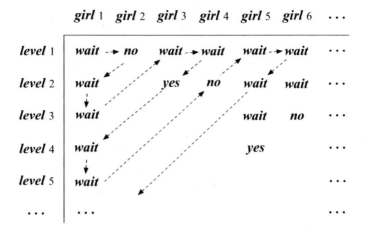

FIGURE 4.1: The table of Martin's adventures. Traversal starts in cell *level* 1 - *girl* 1.

denotes that computation is still ongoing. The arrows indicate the order in which Martin traverses his adventures, approaching the girl with the first incoming arrow and visiting her in later "levels" to check for a decision. Any girl deciding to accept Martin's love will inevitably be reached and seduced in finite time. For example girl 3 will be reached in seven steps since girl 1, approached before, is not willing to make a decision in a short time (and might never decide, i.e., her algorithm might never halt). Then girl 5 will be later seduced, and so on. To sum up, Martin will achieve the seduction of any girl that eventually decide to say **yes**, but he does not forget the others because, in his view

those who do not go after anything but this last level are wretched, primitive men ....

To speed up the process Martin could try to foresee the behavior of the registered women, for example skipping all the useless encounters with girl 1 to consider less reluctant girls. Now acting as an algorithm, he should examine the girls' algorithms as his data. However we know that complete prediction is ultimately impossible. Although exceptionally gifted as a womanizer, Martin knows that his decision power among peers cannot go that far.

4.2 Computational complexity

Confronted with a specific problem one may think of several solutions and design several corresponding algorithms that differ dramatically in their efficiency. A serious comparison must be performed on solid mathematical grounds. Computational complexity theory is aimed at precisely this.

The foundations of this branch of science have been developed under the assumption that computation is carried out by a single computing device with a single block of random access memory. This still constitutes the bulk of the theory, although some adjustments are required for different computing gadgets like, for example, computers containing multiple processing units and various levels of memory. The time required by the execution of an algorithm (or "computation time") is the main parameter for evaluating the quality of the algorithm itself and for comparing one algorithm with another.[6]

Once, both computation time and storage requirement were considered. Today the latter parameter is generally disregarded except for applications conceived for gadgets with a very small amount of memory like cell phones, or for networks where a huge amount of data must be replicated in many locations for fast access. There are various reasons for this. Time cannot be traded for space or vice-versa because one may reside in the same place in different instants, but cannot reside in different places at the same instant. Space can be bought, particularly in the computer world where memory chips have become highly inexpensive; but there is no way of stopping time from passing. We can also observe that memory is used for storing input data, intermediate results arising during the computation, and possibly final results. Even if we assume that input data are pre-loaded, filling the memory with the other items requires time, but using time does not necessarily imply that the memory is involved. So computation time accounts also for the amount of memory used, but not vice-versa.

A novel angle on all this emerged recently when people started to become concerned about the huge amount of energy that computing systems are consuming today. Besides computation time, energy is now becoming another important parameter as we will discuss at the end of this chapter. Note that computation time cannot be measured in seconds without specifying the particular hardware and software to be used, so is instead expressed as an adimensional mathematical function of the size of the input data (or "input size") under the standard assumption that such a time is essentially related to the input size and increases with it. To make this possible very large input sizes are considered and the mere order of magnitude of the growth is taken as being relevant.

Strictly speaking the input size should be expressed as the number of char-

---

[6]Some important considerations in this respect will be treated in a subsequent chapter on parallel and distributed computing where the rules of the game are different.

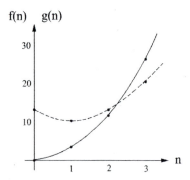

FIGURE 4.2: Comparison between the functions $f(n) = 2n^2 - 4n + 14$ (dashed line) and $g(n) = 3n^2$ (solid line). $f(n)$ is of order $O(n^2)$.

acters needed to describe the input in any reasonable (i.e., at least binary) alphabet. With a standard binary alphabet the input size is then a positive number b of bits, but one can be much less formal taking any integer n proportional to b. That is, the way n grows is of interest, independent of any multiplicative constant. The computation time $T(n)$ is then expressed as a mathematical function of n in a specific order of magnitude, and is strictly significant only for $n \to \infty$. We will content ourselves with reasonably high values of n as always naturally happens dealing with Internet and Web problems. In any case n will always be an integer.

Several orders of magnitude are defined and used in computer science, to express different situations that are unusual in mathematical analysis. We shall refer to the well known "big O" order, used to establish an upper bound to the value of a function $f(n)$ as n grows. Formally $f(n)$ is "of order" $O(g(n))$ if $f(n) \leq c \cdot g(n)$ for a given constant c, and $n \to \infty$. With a certain amout of abuse of the mathematical language it is common to write $f(n) = O(g(n))$. So for instance $f(n) = 2n^2 - 4n + 14$ is of order $O(n^2)$, also written $f(n) = O(n^2)$, since $f(n) \leq c \cdot n^2$ for $c = 3$ and $n \geq 3$ (see Figure 4.2).[7]

The computation time $T(n)$ of an algorithm, expressed in order $O(g(n))$, certifies that the rate of growth of $T(n)$ is limited from above by $g(n)$ for any large value of n. The meaning of this statement, however, is less obvious than it may appear because the computation time may depend not only on n, but also on the specific input. In fact input data of the same size may require different computation times. The expression $T(n) = O(g(n))$ says that the

---

[7]A nice metaphore due to Gadi Landau explains the concept of order of magnitude as follows: a billionaire is always a billionaire if he owns one or nine billion dollars, and remains the same even with some millions more.

upper bound $g(n)$ applies to *all* the input data of size n, and $g(n)$ is said to be the *worst case complexity* of the algorithm.[8]

Following this line we can group the algorithms in different complexity classes, assigning to the same class those having the same or similar order of complexity. The main classes arising in practice contain: functions of *constant* order, whose growth is independent of n; functions of order less than n called *sub-linear*; functions of order n called *linear*; functions of order $n \log n$, for which the base of the logarithm is non influential;[9] functions of order n^2, n^3, etc., collectively called *polynomial*; and functions of order k^n, n^n etc., collectively called *exponential* because n appears as an exponent.

Let us now enter the world of algorithm complexity through the classical problem of *sorting*, or arranging a set of elements in order. So many operations are facilitated when this is done that sorting is generally considered as the problem to start with. Furthermore it is a cute problem that has been studied in great detail. Formally, start from a set A of n distinct elements to be sorted in "increasing order." Strictly speaking, for any two elements of A we should define an ordering relation that can be denoted by $<$, to decide which one comes first. In practice this problem is overcome automatically by the way data are represented in any computer, since all characters of the alphabet at hand are represented in binary. Therefore sorting "words" of any type amounts to comparing the corresponding binary strings, taking in turn these strings as binary numbers and applying a simple numerical comparison. For example in the standard Internet code UTF-8 the letters of the alphabet correspond to progressive binary numbers, so that the word ALFA is indeed "smaller" than BETA. So, for simplicity, we will directly represent all data with integers.

Selecting amongst many, we will consider three sorting algorithms called *Selection Sort*, *Merge Sort*, and *Permutation Sort*. The first two are well known and in use in practical applications. The third is a zany invention of the authors (the reason for being zany will be clear later). Our aim is to compare the three algorithms through the evaluation of their computational complexities in the worst case, as a function of the number n of elements to be sorted. For this purpose arrange these elements in a mono-dimensional array $A[1 : n]$, with $A[k]$ and $A[i : k]$ respectively denoting the element in position k, and the sequence of elements in positions i to k. The basic structure of the three algorithms is shown in Figures 4.3 to 4.5.

Selection Sort (Figure 4.3) should be clear if the reader recalls how a recursive procedure works (e.g., see the recursive rule for tree pre-order traversal in Figure 3.10 of the previous chapter). The computation goes through $n - 1$

---

[8]The word "complexity" has different meanings in different sciences. In the field of algorithms we refer to the (time) complexity *of an algorithm* A as the computation time of A expressed as an adimensional function of n. It is also customary to define the complexity *of a problem* P as a lower bound to the complexity of any algorithm that solves P. This amounts to proving that no computation can do better than a given bound.

[9]As explained in Chapter 3, relation (3.1), logarithms with different bases a, b differ only by the multiplicative constant $\log_a b$.

algorithm SELSORT(A, k)

    if $k > 1$

        find the maximum M in $A[1 : k]$;

        exchange M with $A[k]$;

        SELSORT($A, k - 1$);

FIGURE 4.3: Selection Sort. The computation starts with $k = n$, goes through $n - 1$ cycles with k decreasing from n to 2, and terminates for $k = 1$.

cycles with decreasing k. In each cycle a prefix $A[1 : k]$ of the array is scanned searching for its maximum element M that is then moved to the last position of the prefix; so M finds its final position in the array. M can be determined in different ways, for example starting with $M = -\infty$, comparing its current value with each element of the prefix, and updating the value of M each time a greater element is encountered.

In any case M can be found with $O(k)$ operations in a prefix of length k. Since all the prefixes of length n to 2 are scanned, the total number of operation for finding M in all the cycles is $O(n + (n - 1) + (n - 2) + ... + 2)$, that is $O(n^2)$ because $n + (n-1) + (n-2) + ... + 2 = (n^2 + n - 2)/2$. Furthermore all the other operations required by Selection Sort, as for example exchanging M with $A[k]$, are executed once per cycle. Then the total time is dominated by the time for scanning. Denoting by $T_1(n)$ the computation time of Selection Sort we then have:

$$T_1(n) = O(n^2). \tag{4.5}$$

Merge-Sort (Figure 4.4), a very popular sorting algorithm, is based on the merge of two already sorted subsets of approximately the same size into a unique sorted set. In Figure 4.4 this is done by the procedure whose structure will be now described in words. Suppose we have two stacks of $n/2$ student papers each, sorted by increasing id number. The basic step of MERGE consists of selecting the paper P with minimum id from the top of the two stacks, removing P from its stack so that a new paper appears, and putting P onto an output stack (face down). The step is repeated until one of the input stacks becomes empty. At this point all the papers still in the other stack go onto the output stack without any further comparison because their ids are certainly greater than the ones already in the output. Since each id comparison involves taking one paper off one of the input stacks, a total of at most $n - 1$ comparisons are required to empty both stacks. The computation time $T_M(n)$ of MERGE is then of order $O(n)$.

The computation time $T_2(n)$ of the whole algorithm MERGESORT can now be expressed through the *recurrence relation* $T_2(n) = 2\,T_2(n/2) + T_M(n) + c$, where the term $2\,T_2(n/2)$ accounts for the two recursive calls of the algorithm on subsets of size $n/2$; $T_M(n)$ is the time of MERGE; and c is a constant

algorithm MERGESORT(A, i, j)

   if $i < j$

      $k = \lfloor (i + j)/2 \rfloor$;

      MERGESORT(A, i, k);

      MERGESORT$(A, k + 1, j)$;

      MERGE$(A[i : k], A[k + 1 : j])$;

FIGURE 4.4: Merge Sort. The computation refers to a sub-array $A[i : j]$; starts with $i = 1$, $j = n$; goes through successive recursions of the procedure on two sub-arrays of half length; and merges these sub-arrays into $A[i : j]$ again after they have been recursively sorted.

algorithm PERMSORT(A)

   generate the permutations of A one by one

      for-each permutation

         if $\{A[i] < A[i + 1]$ with $1 \leq i \leq n - 1\}$ **stop**;

FIGURE 4.5: Permutation Sort. As soon as a permutation of A is generated, the algorithm checks whether it is sorted.

time needed for all the other operations as, for example, computing the central position k in $A[i : j]$. Since $T_M(n) = O(n)$, the recurrence relation above can be solved with standard mathematics to obtain:

$$T_2(n) = O(n \log n). \tag{4.6}$$

Hence Merge Sort is definitely better than Selection Sort.[10]

The third algorithm Permutation Sort (Figure 4.5) is not very smart. All the permutations of A are generated one after the other and, for each one of them, the algorithm checks if the elements are in increasing order, stopping when this condition is finally met. This control requires $O(n)$ time for each permutation, but the computation can go on for a long time. In the worst case the algorithm is required to produce all the $n!$ permutations before stopping. The computation time $T_3(n)$ of Permutation Sort is then of order $O(n!n)$. Using Stirling's approximation (see Chapter 2): $n! \sim (2\pi n)^{1/2}(n/e)^n$, we have:

$$T_3(n) = O(n^{3/2}(n/e)^n). \tag{4.7}$$

Therefore Permutation Sort is not only by far the worst algorithm among the three, but is so inefficient to justify a drastic division of the world of algorithms into two big complexity classes: the one of *polynomial* algorithms which includes Selection Sort and Merge Sort; and the one of *exponential* algorithms for Permutation Sort and all the ones whose computation time

---

[10]To study how to solve recurrence relations see, for example, Cormen T.H. *et al.*, 2009.

is an exponential function of n.[11] Some further considerations may better explain the reason for such a distinction.

Assume that the running times of three programs implementing the algorithms above on a given computer can be exactly evaluated, say, in microseconds. Assume further that such times can be expressed as:

$$R_1(n) = n^2,$$
$$R_2(n) = n\, log_2 n, \tag{4.8}$$
$$R_3(n) = n^{3/2}(n/e)^n.$$

We have discarded all the terms of lower order and have assumed that the multiplicative constants are all equal. These simplifications obviously affect the numerical results but not the asymptotic behavior of the functions.

Figure 4.6 reports the values of the functions (4.8) for some values of n, doubling each time. While sorting eight elements requires tens of microseconds in all cases, so the three algorithms are essentially equivalent for small values of n, for increasing n Selection Sort (time R_1) is clearly worst than Merge Sort (time R_2), but Permutation Sort (time R_3) clearly belongs to another world. To get a perception on how absurd is this algorithm, consider that there are "only" about $3.15 \cdot 10^{12}$ microseconds in one year!

It should be also clear that the tremendous growth of $R_3(n)$ is not due to the polynomial term $n^{3/2}$, and even the fact that the base n/e of the exponential term grows with n could be ignored, as the phenomenon is essentially due to the appearance of n at the exponent. For instance a hypothetical sorting algorithm with running time 2^n, even if better than $R_3(n)$, would still require about 10^{26} years to sort 128 elements.

Another consideration has to do with the benefits that one could get in the future using computers of continuously improving speed. For this purpose it is sufficient to compare the two functions n^c and c^n with constant c, as representatives of the polynomial and the exponential worlds. Roughly speaking let a computer $\bar{C}$ of tomorrow be k times faster than a computer C of today, whatever this means. As a first approximation, we may assume that using $\bar{C}$ for a time t corresponds to using C for a time $k\,t$. If we can do something today we will be able to do much more tomorrow, but how can we measure this advantage? More precisely we can ask ourselves: given an algorithm, if a problem of size n can be solved in time t using C, which size $\bar{n}$ will be attainable using $\bar{C}$ for the same time t? This simple question bears an interesting insight in computational complexity.

For the function n^c we have:

$$n^c = t \quad \text{and} \quad \bar{n}^c = k\,t, \quad \Rightarrow \quad \bar{n} = k^{1/c}\,t. \tag{4.9}$$

---

[11]In fact the computation time of Merge Sort is $O(n \log n)$ which is not a polynomial function. However this function can be bounded from above for example by $O(n^{1.5})$ and assigned to the polynomial class.

n :	8	16	32	64	128
R_1 :	64	256	1024	4096	16384
R_2 :	24	64	160	384	896
R_3 :	127	$1.33 \cdot 10^{11}$	$3.36 \cdot 10^{36}$	$3.27 \cdot 10^{90}$	$1.99 \cdot 10^{214}$

FIGURE 4.6: Running times in microseconds of an implementation of the three sorting algorithms, for some values of the number n of elements to be sorted (functions (4.8)).

That is, the better a polynomial algorithm is (i.e., the lower is the exponent c), the more it benefits from the increase of speed of the computational media. Linear algorithms ($c = 1$) can solve problems whose size $\bar{n}$ is k times larger than n; for quadratic algorithms ($c = 2$) this benefit reduces to $\sqrt{k}$; and so on. For the function c^n, instead, be prepared to face a very negative surprise. We have:

$$c^n = t \text{ and } c^{\bar{n}} = k\,t, \Rightarrow \bar{n} = n + \log_c k. \qquad (4.10)$$

In this case the benefit is limited to the *additional* term $\log_c k$, where the increment k of computer speed is also strongly reduced by a logarithmic function. For the hypothetical sorting algorithm running time 2^n, a computer 1024 times faster would sort only $\log_2 1024 = 10$ more elements in the same time period. So in 10^{26} years it could sort 138 elements instead of 128. In a later chapter we will see that the same limitation applies to parallel computers no matter how many processing elements they contain.

From all this we conclude that exponential algorithms are useless, now and in the future, except for extremely small values of the input size n. Unfortunately some problems are intrinsically exponential. A trivial example is an algorithm which generates an output whose size is exponential in n, like the one we have already encountered for finding all the permutations of a given set. But the situation is much subtler as we shall see below.

4.3 Searching: a basic Internet problem

A basic request of almost all Internet users is to retrieve information connected to one or more freely specified keywords. These keywords must be stored somewhere on the net, together with pointers that indicate where the related information can be found. A whole chapter of this book will be devoted

algorithm BINSEARCH(e, A, l, r)

    if $l > r$ **answer**:

                e *does not belong to* A;

                the predecessor of e *is in position* r;

                the successor of e *is in position* l;

    $k = \lfloor (l + r)/2 \rfloor$;

    if $A[k] = e$ **answer**: e *is in position* k;

    if $A[k] > e$ BINSEARCH$(e, A, l, c - 1)$

        else BINSEARCH$(e, A, c + 1, r)$;

FIGURE 4.7: Binary search of e in $A[l : r]$. The computation starts with $l = 1$, $r = n$; goes through successive recursions of the procedure on a sub-array of approximately half length; and terminates when e is found (in position k), or the searched sub-array is empty. The command **answer** implies that the execution terminates.

to explaining how this is done. From an algorithmic point of view, the search problem in general is preeminent for us.

In principle search is formulated as the problem of looking for a given element e in a set A of n elements. From our discussion of the representation of information inside a computer, we know that the elements are actually binary strings and may be always treated as integers, although further attention will be given to them when dealing with extremely large sets. Just as for sorting, the set can be stored in a mono-dimensional $A[1 : n]$ with the initial assumption that, given an integer i between 1 and n, the element $A[i]$ can be accessed in constant time. The required output can be the position of e in $A[1 : n]$ if $e \in A$; or the indication that $e \notin A$; or, in the latter case, we can be more demanding and ask for the element of A that is "closest" to e. Dealing with integers this request may be formulated as giving the predecessor, or the successor, of e in the sorted list of all the elements of A. And in fact the search can be directed quickly towards the position that element e should occupy if the array $A[1 : n]$ has been previously sorted, say in ascending order.

The process is similar to searching for a word in a dictionary, and this is in fact what happens on the Web. In a manual human search the initial letter of the word directs the reader towards a specific portion of the book, and the next steps are refined over time depending on the words encountered, according to a largely heuristic mechanism. The computer counterpart does not use any special knowledge of the frequency or the distribution of the letters, but can do something that humans cannot do, namely, computing specific positions in the array in a very short time. Despite these differences, the computation time is roughly the same. The resulting algorithm is the very well known *Binary Search*, given with some non-standard additions in Figure 4.7.

The functioning of algorithm BINSEARCH should be fairly obvious. Starting with the whole set $A[1 : n]$, the central position k is computed, and e is compared with the element $A[k]$ encountered there. If the two elements are different and $A[k] > e$, the search proceeds recursively in the left half of the array that hosts all the elements smaller than $A[k]$. If $A[k] < e$ the search proceeds in the other half of the array. If the sub-array thus determined is empty (condition $l > r$, i.e., the left limit has become greater than the right limit) e is not in the set and its predecessor and successor are immediately found in positions r and l as the reader may verify with a simple thought.

An example of application of algorithm BINSEARCH to a set of $n = 12$ elements and $e = 13$ is shown in Figure 4.8. The search touches the elements 21-5-8-13 (note that when a sub-array has an even number of elements its central position is computed as the right extreme of the left half). A search for $e = 11$ which is not contained in the set would follow the same steps and terminate after the comparison between the $e = 11$ and $A[5] = 13$ with the procedure call: BINSEARCH(11,A,5,4), and the answer that element 11 is not in the set and its predecessor and successor are in position 4 and 5, respectively.

The computation time of Binary Search in the worst case applies when the size of the final sub-array is reduced to one for a successful search (e.g, as in the example of Figure 4.8) or to zero for an unsuccessful search. Each cycle of the algorithm requires a constant number of operations followed by a recursive call on a sub-array whose size approximately halves, so the total number of cycles is approximately $\log_2 n$ and the computation time is of order $O(\log n)$.[12] This is a distinct improvement over the naive strategy of scanning all the elements of A in $O(n)$ time until the one that matches with e (if any) is found.

The computation time of Binary Search does not take into account the $O(n \log n)$ time needed to sort the set A in advance, e.g., using MERGE-SORT. Then Binary Search is advantageous over a naive scanning only if many different searches have to be done on the same sorted set A. The point of balance occurs for $O(\log n)$ searches: on the one hand, we can repeat a sequential scanning $\log n$ times, on the other hand we can sort A and then repeat a Binary Search $\log n$ times, ending up with a total of $O(n \log n)$ time in both cases. For a greater number of searches, as for example in Web operations, Binary Search definitely prevails.

Once a set A is sorted, searching for an element is just one of several operations that can be done very efficiently. Assume for example that k elements $e_1, ..., e_k$ must be retrieved in A, where k is a (small) constant. For example, $e_1, ..., e_k$ are keywords contained in one or more Web queries, and A is a dictionary containing all possible keywords, as will be described in a later chapter. We can assume that $e_1, ..., e_k$ are sorted since this can be done in constant time. Retrieving these elements one after the other would require $O(k \log n)$

---

[12]Recall that as a number $n = 2^s$ can be obtained starting from 1 and doubling the number $s = \log_2 n$ times, so the number 1 is obtained starting from n and halving the number $s = \log_2 n$ times.

	1	2	3	4	5	6	7	8	9	10	11	12
A :	2	3	5̇	8	13	**21**	24	29	35	36	42	45
	2	3	**5**	8	13	21	24	29	35	36	42	45
	2	3	5	**8**	13	21	24	29	35	36	42	45
	2	3	5	8	**13**	21	24	29	35	36	42	45

FIGURE 4.8: Consecutive calls of the algorithm BINSEARCH for $e = 13$. The elements reached at each step are in boldface.

time if Binary Search is applied directly. If e_1 and e_k are instead retrieved first, all the other elements can be searched in the sub-array $A[l : r]$ whose bounds are the positions of the successor of e_1, and of the predecessor of e_k, respectively. And a further refinement of the bounds can be done if e_2 and e_{k-1} are retrieved next, etc. Although the worst case time still is $O(k \log n)$, a great reduction is attained if the bounds l, r in one of the pairs above are close to one another.

Another important application is finding all the common elements of two sorted sets A, B. In a Web search context A and B may be two sets of pages each containing a given keyword, and we are interested in the pages that contain both of them. Now applying Binary Search may or may not be advisable depending on the relative sizes n_A, n_B of the two sets. If, say, $n_A \ll n_B$, a Binary Search in B is applied to each element of A in a total time $O(n_A \cdot \log n_B)$. If instead $n_A \simeq n_B \simeq n$ the above approach would require $O(n \log n)$ time while the search can be done in linear time with the concurrent scanning algorithm DOUBLESEARCH shown in Figure 4.9.

The two sets are scanned advancing the pointers i, j in unitary steps. The computation time is proportional to $n_A + n_B$, that is $O(n)$. The reader should examine the algorithm carefully because the element comparisons are not totally obvious. In particular a terminator must be added to both sets to ensure proper access to the arrays in the last two steps.

The advantage in using Binary Search relies on the assumption that the elements of A do not change for a long enough time to allow many searches to be completed on the same sorted array. Many files, however, change dynamically and, if one element is inserted or deleted, rearranging the array requires $O(n)$ time and all the advantages disappear. This problem can be overcome if the elements of A are identified with the nodes of a binary *search tree* where, by definition, the nodes contained in the left subtree of any node x are smaller than x and those in the right subtree of x are greater than x. Figure 4.10 shows a search tree for the set A of Figure 4.8: node 24 is in the right subtree of 21 and in the left subtree of 35, and indeed we have $21 < 24 < 35$. Note that many other search trees would be consistent with the same set, depending on the element chosen for the root and, recursively, for the roots of all the

algorithm DOUBLESEARCH(A, B)

$\quad i = 1; j = 1;$

$\quad$ **while** $\{(i \leq n_A) \text{ and } (j \leq n_B)\}$

$\quad\quad$ **if** $A[i] = B[j]$

$\quad\quad\quad$ **put** $A[i]$ *in the output*; $i = i + 1; j = j + 1;$

$\quad\quad$ **if** $A[i] < B[j]$ $\quad i = i + 1;$

$\quad\quad$ **if** $A[i] > B[j]$ $\quad j = j + 1;$

FIGURE 4.9: Retrieving common elements in two sorted arrays $A[1 : n_A]$, $B[1 : n_B]$. To ensure proper termination, a special symbol, say \$, is put at the end of both arrays in positions $n_A + 1, n_B + 1$.

subtrees. So, many search trees correspond to the same sorted array, but only one array corresponds to all such trees.

It is not difficult to see that the search for an element e in a tree T is very similar to the Binary Search for e in the corresponding array. e is initially compared with the root r of T, which has the same role of the central element of the array. In the case of a match, the search terminates successfully, otherwise the result of the comparison directs the next step to one of the two subtrees of r where the search is repeated recursively. In fact if $e < r$ (respectively, $e > r$) node e can only be found in the left subtree (respectively, in the right subtree) of r. So, as for Binary Search, a significant portion of the set is disregarded at each step. The search then consists of a sequence of comparisons along a path that starts at the root of T and ends successfully in node e, or unsuccessfully in an empty subtree. In the tree of Figure 4.10 the search for 13 passes through the nodes 21-5-8. The search for 12, which is not in the tree, follows the same path up to the empty left subtree of 13.

The computation time of tree search in the worst case is proportional to the number of nodes in the path between the root and the farthest leaf. Therefore in the unfortunate case that the tree degenerates into a chain this time is $O(n)$. To see if this can actually happen, consider again that these trees have been introduced for inserting and deleting nodes efficiently. Insertion of a new node x is actually very easy because it amounts to searching x in the tree even if one knows that it is not there. This leads to an empty position where x is allocated (see the insertion of 12 in the tree of Figure 4.10). Deletion is slightly more complicated and is omitted here. In any case search, insertion, and deletion of a node require traversing the tree up to a leaf, therefore it is crucial that the leaves are not too far away from the root.

As the reader probably suspects, the best situation arises when all the leaves are maintained at a distance $O(\log n)$ from the root, so that tree search and Binary Search require the same time for each element. However this is not an easy task as the tree changes. It is sufficient to say here that, taking all search trees corresponding to a given set A, the length of the path from

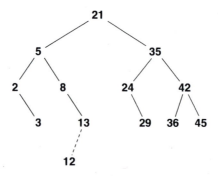

FIGURE 4.10: A binary search tree hosting the set A of Figure 4.8, and the insertion of a new node 12.

the root to the farthest leaf is indeed $O(\log n)$ on the average, so the expected value of the computation time of tree search is of that order even in the worst case.[13]

A last question is related to the amount of information on the set A that a binary search tree actually contains, besides the values of its elements. The answer is that the tree holds the set totally sorted, albeit indirectly, and the ordering can be retrieved in time $O(n)$ by an *in-order traversal* of the tree. Note that this is the time to output the set, i.e., no extra time is required for sorting.

Contrary to the pre-order traversal rule discussed in the Chapter 3, in the in-order traversal, each node is visited between the nodes of its left subtree and those of its right subtree, see Figure 4.11. If the algorithm is applied to a binary search tree, each node x is reached after the nodes smaller than x and before the nodes greater than x, so all the nodes are reached in a sorted sequence. On the tree of Figure 4.10, the algorithm is first called on the root 21; then is recursively called on its left child 5; and again on the left child 2 of 5. Since 2 has no left child, 2 is placed in the output; then the algorithm is recursively called on the right child 3 of 2, etc. The sorted set A finally emerges.

---

[13]A complete discussion of search trees can be found in any book on algorithms and data structures, see for example Cormen T.H. *et al.*, *op. cit.*.

algorithm IN-TRAV(x)

   if $\{x$ has a left child $c_l\}$ IN-TRAV(c_l);

   output (x);

   if $\{x$ has a right child $c_r\}$ IN-TRAV(c_r);

FIGURE 4.11: A recursive algorithm for in-order traversal of a binary tree. Computation is triggered by letting $x = root$.

4.4 Lower bounds

Once we have seen how the computation time of an algorithm is evaluated, it is natural to ask if one can do better. That is, if a problem P is solved with algorithm A in time $O(f(n))$, might a faster algorithm exist for P? If not, it is useless to keep on trying. Formally the aim is to find the intrinsic complexity of P, that is, a lower bound on the computation time that no algorithm for P may overcome. For some problems, finding such a bound is easy; for others, it is difficult; for still others a certain bound is suspected but is extremely hard to prove.

As computation times are evaluated in order of magnitude, a new order must be defined to express lower bounds. In the science of algorithms this is the order "big Omega," in a sense symmetrical to "big O." Formally $f(n)$ is of order $\Omega(g(n))$ if $f(n) \geq c \cdot g(n)$ for constant c and $n \to \infty$. That is, for large values of n the function $f(n)$ grows *at least* as $g(n)$. Of course trivial lower bounds to the computation time can always be established. For example a lower bound $\Omega(k)$, with k a positive constant, applies to any problem since all algorithms must perform at least one operation when called. What may be difficult is finding a *significant* lower bound for a given problem P or, ideally, an absolute bound that no algorithm for P can beat. If reached, this bound is like a wall that blocks the passage to better algorithms.

There is a certain symmetry to the situation discussed before. Better and better algorithms for P are found, and each time we retain the one with the lowest computation time, while higher and higher lower bounds for P are sought for, and each time we retain the highest. On one side new algorithmic schemes must be invented to attain a better performance, and on the other side new properties of the problem must be discovered to prove that a certain number of steps must be necessarily performed. Recall that in all cases we work in order of magnitude. When the computation time of the best algorithm meets the highest lower bound we can declare that the algorithm is *optimal* and P is *computationally closed*. Unfortunately this seldom happens, and we must leave the two bounds open for further improvements. For many problems, lower bounds have received little attention and only trivial bounds are known.

Let us take as an example two fundamental problems in the field of algorithms: sorting a set of n distinct elements, considered in section 4.2; and multiplying two bi-dimensional arrays of $n \times n$ elements, discussed in the next Chapter 6 in relation to finding paths in a network. Three algorithms have been considered for sorting, the best of which is Merge Sort that requires $O(n \log n)$ time (relation 4.6). The question is whether one can do better. A trivial lower bound for sorting n elements is $\Omega(n)$ because, to put all the elemens in order, each one of them must be examined at least once. Between $\Omega(n)$ and $O(n \log n)$ there is room for improvement, in principle from both sides. We prove now that this case is unusually favorable and the lower bound can be raised significantly.

Two facts are basic. First, sorting a set A amounts to building the unique sorted permutation of its elements out of all the $n!$ permutations of A. Second, any sorting algorithm must compare elements pairwise, besides doing many other operations. So if we can establish a lower bound to the number of comparisons needed to identify a permutation out of $n!$, then this is also a lower bound for the whole algorithm. Let us start comparing two arbitrary elements x, y. The best we can hope is that the set of all permutations is split into two non overlapping subsets where the solution may be found, one for $x < y$ and the other for $x > y$. Furthermore these two subsets must contain the same number of permutations, otherwise, in the worst case, the solution would lie in the bigger one. Then, at the best, the set of all permutations is recursively halved at each comparison until a set with exactly one permutation (the sorted set) remains. To do so $\log_2 n!$ comparisons are needed, and applying Stirling's approximation we have:

$$\log_2 n! \sim \log_2((2\pi n)^{1/2}(n/e)^n)$$
$$= (1/2)\log_2(2\pi n) + n\log_2 n - n\log_2 e = \Omega(n \log n)$$

where the second term in the addition prevails. Then $\Omega(n \log n)$ is a lower bound to sorting. Since this bound matches the computation time of Merge Sort we can declare that Merge Sort is optimal (in order of magnitude) and the problem is computationally closed.

The problem of matrix multiplication, however, seems at first glance fairly straightforward, but turns out to be computationally messy. We will not reproduce here a discussion of its role and solution, relying on some recollection that the reader may have from high-school linear algebra. A trivial lower bound of $\Omega(n^2)$ arises from the need to examine all the $2n^2$ elements of the two arrays to be multiplied, but essentially no higher lower bound is known. On the other hand many algorithms have been proposed for its solution, with a computation time that varies from a straightforward $O(n^3)$ scheme to very sophisticated $O(n^k)$ schemes with $2 < k < 3$. In spite of its fundamental importance, this problem is still computationally open.

In the next section we will meet new facts on lower bounds that are really surprising.

4.5   A world of exponential problems

In our discussion of computational complexity we mentioned that there are problems whose solution requires exponential time. This is obvious if an exponential lower bound can be found, as for example when asking for an output of exponential size (i.e., printing all the permutations of a set). There are other problems, however, naturally defined and apparently harmless, for which the only known algorithms require exponential time, but no polynomial lower bounds have been found. The gap between lower and upper bound is huge. Nobody is able to solve these problems efficiently, i.e., in polynomial time, but neither can anybody prove that this is impossible. Everybody, however, is practically convinced that this is the case.

Noneless to say, a wealth of interesting and useful problems fall into this set! To explain what is behind this challenging situation we start with a story of pure fiction.

Story. A criminal group decides to use the Web to exchange secret information and forbidden material. They decide that the best way to avoid attracting attention is to join a legitimate social network, establish relationships inside it and exchange encrypted secret messages with each other. After a while, some rumors of the illegal activity of the group reach the police. Being unable to decode the contents of the messages, but wishing to identify the members of the group, the police decide to analyze the graph of the social network to identify communities with common interests. The task is assigned to the youngest officer.

The social network includes one thousand people, and the criminal group is estimated to consist of ten of them. Therefore the officer has to find whether a complete sub-graph of a size ten, or a 10-*clique* in graph terminology, is contained in the network. Being an expert in computer programming, he decides to organize the search systematically, but has no better idea than to consider all possible groups of ten nodes in turn and verify whether all possible connections between pairs of nodes are present in each group. If so, the group is a 10-clique, and the participants merit intensive screening. For this purpose he assigns a number from 1 to 1000 to each node of the graph and writes a program to generate all groups of ten nodes starting with (1, 2, 3, ..., 9, 10), checking whether the group is a clique, and then increasing the value of each element of the group in unitary steps starting from the last ones.

Therefore the groups should be generated in the order: (1, 2, ..., 9, 10), (1, 2, ..., 9, 11), ..., (991, 992, ...,1000), but they are so many that, after some hours, the program has not yet output a final result when the lieutenant arrives. We let you imagine his fury at such an apparent inefficiency and the face of the worried programmer: neither of the two knows that no better algorithm than that used by the young officer essentially exists, namely listing one after the

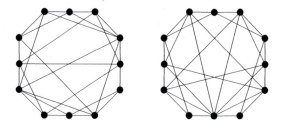

FIGURE 4.12: Searching for 4-cliques in two graphs of twelve nodes. One graph has no such cliques; the other has two of them.

other all the possible groups of ten to decide whether a 10-clique is hiding in the graph.

To understand the meaning of the story we must consider how to build an algorithm from a more general point of view. The number of groups of ten elements out of one thousand equals the binomial coefficient $\binom{1000}{10} \simeq 2.64 \times 10^{24}$. Recalling the definition of binomial coefficients, and applying Stirling's approximation once again, it can be easily seen that $(n/k)^k < \binom{n}{k} < n^k$, that is the function is exponential in k and becomes exponential in n when k grows linearly with n. In our story we have $n = 1000$ and $k = n/100$, so the above inequalities indicate that the value of $\binom{n}{k}$ lies between 10^{20} and 10^{30}. Unfortunately no efficient strategy is known for discovering k-clique in a graph or reporting that there is no such sub-graph. This statement must be taken in a very general sense, because for particular graphs the answer may be easy or even obvious: e.g., since each node of a k-clique must be connected to $k-1$ other nodes in the clique, only the nodes of the original graph with degree at least $k - 1$ must be considered. In general, however, the only known way of approaching this problem is to examine all possible groups of k nodes, as the police officer did, or applying other equivalent methods based on complete enumeration. The reader may appreciate how difficult the search is by looking for 4-cliques in the two graphs of Figure 4.12.

Problems for which only exponential time algorithms are known are called *intractable* to indicate that in practice they cannot be solved by a computer as the input size increases. In fact for many intractable problems there is no definitive proof that a polynomial time algorithm does not exist, so they must be considered intractable until the (albeit unlikely) discovery of such an algorithm. Let us look at this situation more deeply.

After the algorithmic community began to be aware of the existence of intractable problems and of some relations between them, deep theoretical studies started in the 1970s with a famous result often called the Cook-Levine theorem, independently proved by the American-Canadian Stephen Cook and the Russian Leonid Levin. The theorem is far too specialized to be discussed here. A fundamental consequence, however, was to show the existence of a large class of problems tightly linked to one another by an unexpected relationship: if we were able to solve even only one of them in polynomial time, all the problems of the class would also be solvable in polynomial time. On the other hand if we were able to find an exponential lower bound for one of them, all the problems of the class would provably require exponential time. These problems are called *NP-hard*:[14] they have in common a logical structure responsible for their difficulty.

Highly diverse problems belong to the class NP-hard. From finding solutions to particular algebraic equations; to replicating Web pages on different servers; to building shortest routing paths in a network; to, of course, graph problems like finding cliques, or finding Hamiltonian cycles as discussed in Chapter 2. However NP-hard problems and the ones like the existence of a k-clique have an interesting property in common: it is difficult to find a solution but is easy to verify that a certain object is indeed a solution.

For these problems finding a solution appears exponentially difficult; but, once a candidate solution is known it can be tested in polynomial time. Then verifying that a group of k nodes in a graph is a k-clique amounts to control that all the nodes in the group are pairwise connected, an operation requiring $O(k^2)$ time because $k(k-1)/2$ is the number of all possible connections in the group.

This ability to test a candidate solution in polynomial time is in fact common to all NP-hard problems, and may be taken as a part of their definition. Informally these are problems that we are not able to solve in polynomial time, but for which the soundness of a proposed solution can be checked in polynomial time.

It must be observed, however, that these problems are hard to solve in the "worst case," that is, when their data occur in a most unfavorable combination. An exponential time algorithm would solve them in all cases, but a polynomial time algorithm could work in particular situations. If we take the computers of a local area network as being the nodes of a graph, this graph obviously has a Hamiltonian cycle if the network is a ring, and there is no cycle if the network is a tree. Both these configurations can be checked for in linear time, so if one of these two cases occurs the Hamiltonian cycle problem is solved immediately.[15] But an algorithm valid for all possible graphs is exponential as far as we know today. Algorithmic techniques to solve NP-hard

---

[14]It would be too long to motivate this name. We simply recall that NP stands for *Nondeterministic Polynomial* to indicate that these problems are polynomially solvable only in a non-deterministic paradigm of computation that exists only as a theoretical abstraction. A related word is *NP-complete* to denote a class of problems where only the existence of a solution with given properties is in question (i.e., the existence of a k-clique) without actually having to report the solution itself. The two classes are subtly related, see the bibliographic notes.

[15]More interestingly, if the network is an $n \times m$ two-dimensional grid where each node is connected to its four neighbors (or to its three, or two, neighbors for nodes on the grid edges, or corners), a Hamiltonian cycle exists if and only if n and/or m is even. The reader can spend a few moments to prove this statement personally, treating the grid as a black and white chessboard.

problems in polynomial time whenever their data allow for this are still to be invented.

In conclusion, we underline once again that many interesting problems are NP-hard, and the hope of finding efficient algorithms for solving them in all cases is actually very low. Hence the next question is: how can we solve these problems in practice, when the size of the input is too big to use an exponential algorithm? A wise answer is: when the optimal solution is too difficult to find, an almost optimal solution will be good enough. In some cases we can use an approximation algorithm for finding a solution not too far from optimal. In other cases we can use a randomized algorithm (see the next chapter), thus finding an exact solution with an extremely low probability of error.

4.6   Computation goes green

Up to this point in our discussion, and up until quite recently in the world of computers, the time required by an algorithm has been the basic parameter to judge its quality. With the massive spread of information technology (IT), however, a new concern has been rapidly growing. Namely, how much energy this technology requires; or, for those concerned with ecology, how much does it contribute to global warming. In the field of algorithms, however, these studies are still in their infancy, so we will just mention where research is going.

Early studies of the 1980s were mainly related to the energy consumption in VLSI technology.[16] In the following decades the focus was shifted to computation in a more general sense until, in April 2009, the National Science Foundation of the U.S.A. sponsored a general meeting on what is now called "energy-proportional computing" to indicate the functioning of systems that balance performance and energy consumption. For algorithm designers this is just a start.

The first observation is that most of the energy used by any IT system goes into heat dissipation. This has an immediate impact in large computer network infrastructures where a huge concentration of hardware inevitably requires expensive cooling and other energy consuming services. Great attention has then been directed to the energy efficient organization of such infrastructures, with almost no relation to computational strategies. The major studies in computer science span from the consolidated areas of solid state and archi-

---

[16]Remarkable studies led to a general framework for measuring the "switching energy" of a logical circuit, as suggested in the seminal paper: Kissin, G. Upper and Lower Bounds on Switching Energy in VLSI. *Journal of the ACM*, Vol. 38 (1) 1991. A work that some people seem to have forgotten.

tectural design, to the emerging area of energy efficient algorithms. A topic now considered with attention, although the problem is not at all simple and still lacks some concrete foundations.

Clearly there is a strong relation between the computation time of an algorithm and the energy needed for it, since the energy consumption depends on the number of instructions and memory accesses done; but the two parameters are not simply proportional because the power required by CPUs and by memories of different types depend on the mode of operation. CPUs can operate at different speeds, with a drastic reduction of energy consumption at lower speeds, and may consume almost no energy when they are idle. A well known rule of CMOS technology, the most widely used today, states that the power p absorbed by a circuit is approximatey poportional to the cube of the speed s at which the circuit operates, i.e., $p \propto s^3$. Then, if we consider a computation fully done in CMOS and requiring time t, the energy $w = pt$ consumed by the circuit can be approximately expressed as: $w \propto s^3 t$. Increasing the speed from s to $s' = ks$, with $k > 1$, hence reducing the computation time from t to $t' = t/k$, implies increasing the energy to:

$$w' \propto (ks)^3 t/k = wk^2. \tag{4.11}$$

So working at a high speed is very costly in terms of energy, as experienced by anyone who has used a battery operated laptop. Reducing the CPU time by k implies k^2 more heat dissipation.

Relation (4.11) is merely indicative of what can occur, because executing an algorithm is exceedingly more complicated than using a CPU at constant speed. First any computation needs memory and the energy required depends on the storage system used. Solid state memories like DRAMs consume sizable energy even when they are not accessed. Disk drives consume no energy in a sleep state, but bringing them back to action may imply a latency access of 1,000 times more than usual, leaving other parts of the computing system consuming energy for nothing.

One relatively well studied case involves a CPU that may operate in different states with different power consumptions. As passing from a lower to a higher power state requires considerable energy, the problem is establishing the best instance in which to perform a state transition. The decision is made difficult by the practical impossibility of knowing, at any instant, the timing of future events that may depend on memory accesses or other system actions. The problem has then been approached for *online* operations where decisions have to be taken on the spot, either deterministically or relying on a probability distribution of the state transitions. This approach is very technical and is likely to see strong advancements in the near future. We refer the reader to the bibliographic notes.

Formally there is a need for an abstract machine model that reflects the structure of computing systems and their memory hierarchy, as a basis for a new formulation of computational complexity that takes into account the interacting parameters of time and energy. Although challenging, this could lead to a continuous spectrum of evaluation, spanning between two limits: on one side a performance-based view where the computation time is minimized within a maximum energy allowance; on the other side a heat-aware view where the energy consumption is minimized within a maximum time bound. All evaluated in terms of a proper order of magnitude.

Algorithmics scholars must be happy as new theory is needed. Algorithms must go green: the game is fully opened.

Bibliographic notes

The references for section 4.1 can be found in books of computability and mathematical logic. Since these topics are a little out of the mainstream, let us simply mention a small and very pleasant book: Crossley, J.N. *et al.* 1972. *What is Mathematical Logic?* Oxford University Press.

From section 4.2 to 4.5, the reference books are the ones on algorithms and data structures. A very nice one, written with uncommon verve, clarity, and imagination is: Harel, D. 2004. *Algorithmics: The Spirit of Computing*. Third Edition. Addison-Wesley. A classic textbook is: Cormen T.H., C.E. Leiserson, R.L. Rivest, and C. Stein. 2009. *Introduction to Algorithms*, Third Edition. M.I.T. Press.

Readers interested in going even more deeply into the field of intractable problems can refer to the classic (and excellent) book: Garey, M.R. and D. S. Johnson. 1979. *Computer and Intractability: A Guide to the Theory for NP-Completeness*. W.H. Freeman and Co. The book spans from a rigorous theoretical treatment of NP-completeness to a very long list of NP-complete and NP-hard problems grouped by similarity of subjects.

The field of energy-proportional computing is developing very fast, so any specific bibliographic reference may quickly become obsolete. Frequent updates on this subject appear in *Computer*, a journal of the IEEE Computer Society, that publishes scholarly yet simply accessible articles on the main issues of computer science. A very good review on the algorithmic aspects of the field is contained in the paper: Albers, S. 2010. Energy Efficient Algorithms. *Communications of the ACM.* Vol. 53, No. 5, pp. 86-96. Association for Computing Machinery.

As a curiosity, the story of Martin (section 4.1) was first applied to decidability in the Italian book: Luccio, F. and L. Pagli. 1999. *Algoritmi, divinità e gente comune.* (Algorithms, Divinities, and Ordinary People). ETS Editions.

Finally it is worth mentioning the series of international conferences entitled *Fun with Algorithms* (*FUN* for short) that are held in Italy every three years, aimed at presenting rigorous yet witty and amusing papers in the algorithmic area at large. The best papers have been put together in special issues of leading scientific journals, the last ones in *Theory of Computing Systems* (*TOCS*) published by Springer.

Chapter 5

A world of randomness

How the Marquis de Laplace had a strong intuition that was proved correct a century and a half later, and how many other things that happened afterwards were to leave a clear imprint in the digitized world.

The study of mathematical chance is a relatively recent development. Democritus, a Greek philosopher, is often credited as being the earliest scholar of random phenomena for his assertion that the universe was born from the collisions of tiny atoms agitating in chaos. Lucretius, a Roman poet and natural scientist, went much further, giving an explanation of the restless motion of the particles of dust in a sunlight ray that anticipated, with astonishing precision, Einstein's theory of Brownian motion.[1] However none of the ancient sages seem to have thought deeply about the mathematical rules of chance. Random phenomena were appreciated in relation to gambling, but it was not until the sixteenth century that a real treatise on the theory of chance appeared. No wonder that the author, a weirdo Italian with multiform interests, was a hardened gambler. His name was Gerolamo Cardano, a mathematician and physician, with a body disfigured by one of the recurrent epidemics of Black Death that tormented Europe in those times, and a consuming addiction to gambling.

In Cardano's book *Liber de ludo alee* (The Book of Games of Chance), a primitive version of the theory of probability was laid down, along the lines followed by nearly all the other mathematicians until the twentieth century when these studies were given new, stronger foundations. What Cardano established is something that may sound very familiar, that is, the probability of a favorable event equals the number of *all favorable events* divided by the number of *all possible events*. So for example if you bet on an outcome of 2 on a six sided die, your chance of winning is 1/6. This definition lasted for centuries, but requires that all events occur with the same probability (i.e.,

---

[1]Lucretius lived in the first century B.C. Many assertions contained in his poem *De Rerum Natura* (On the Nature of Things) have proved correct many centuries later, although they are not at all intuitive. For example he stated that all bodies would fall with equal acceleration in the vacuum, although he could not conduct any experiment to validate his theory. All this is described in elegant verse.

that probability is defined in circular fashion, in terms of itself). The reason why most textbooks place the official birth of probability one century later may be due to the fact that Cardano, beaten down by fortune and looked down upon by the church, and possibly also unwilling to teach the rules of the game to competing gamblers, kept the book for himself, and it remained unpublished for a century.

Before we get into the mathematics, however, let us look into the art of divination, the other popular field where chance is present. This art can be observed in all cultures in ancient times and still today. The interaction with a supernatural realm is generally put in the hands of a seer (at times a cheat), whose soul travels through the world of spirits to interpret their signs and, if asked to bring guidance to a needy soul, acts as a poetic on-stage psychotherapist. Of course one may deny that a source of wisdom outside of the physical world exists at all, and label these practices as superstitious. But it is difficult to ignore the emotional charge of some oracular incantations, like those of the I Ching, encountered in Chapter 3, or the invocation of a Nigerian babalawo that asks even the trees to remain silent while the spirits are approaching: [2]

Let the Apá be told
to make no sound.
Let the Oró tree be told
to remain silent.
Let the Àsúnrinnikànbéjù tree be told
to cease from talking of his prowess in the jungle.

The quality of the interaction between the seer and the higher powers is the first of three independent components that may occur in any intensity, locating a divination rite as a geometrical point in placed in a 3 dimensional Cartesian space. The second component, we are sorry to say, is fraud, whose amount is larger or smaller than expected depending if one looks upon divination with sympathy or suspicion. We belong to the former category and believe that diviners are for the most part honest, but it is undeniable that throughout history there have been great cheats among them. The Greeks and the Romans deeply respected their oracles whose answers were so ambiguous as to allow any interpretation afterwards, but had many doubts on the honesty of the priests in charge of interpreting the signs of the gods. [3] In several Arab countries there are seers credited with being able to discover the perpetrator of a robbery by visualising the scene in their mind. But often they do

---

[2] Taken from: McClelland, E.M. 1982. *The Cult of Ifá among the Yoruba.* Ethnographica, London. Although it is practically impossible to render in translation the emotion conveyed by a sentence in a strongly tonal language like Yoruba, such invocations are nonetheless impressive and moving.

[3] For example Cicero wrote the book *De Divinatione* (On Divination) to support religion that had an important role in maintaining social order. Still the book is filled with provoking and witty comments on the priests practicing divination.

it for money, so a cynical Yemenite proverb goes: "What the thief missed the astrologer has carried off."[4]

The third ineludible component of divination is chance. Every rite starts with the observation of a random event that occurs spontaneously, like the flight of birds from the Orient or from the Occident in the Roman tradition, or is induced deliberately, like the flight of a bird chased away from its nest by a stone throwing, as in the Arab *zagr*. But in any case the event must be random in a pure mathematical sense, as will be explained below, and even the diviner must not be able to predict its outcome. There is no slight on the power of the seer if the event is truly governed by supernatural forces, but the independence of the components of chance and fraud may become questionable if a cheating diviner tried to influence the outcome. This could possibly happen (but we trust that it never has happened) in the two most perfect divination systems that have been ever set up, that is, the Chinese I Ching and the Nigerian Ifá, where the starting event is respectively induced by flipping coins or throwing half-nuts.[5]

Due to the imperfection of these mechanical devices, or the asymmetry of the diviners hand, the different outcomes of the experiment may occur with unbalanced frequencies even in the absolute absence of fraud. In computer science terms we are in front of a generator of *pseudo-randomness*. This is not at all a problem for divination where the right answers are determined by supernatural spirits regardless of trivial human inventions like probabilities. However, it does indicate that Cardano's definitions must be refined, as they were in the mid twentieth century by Andrej Kolmogorov, a Russian genius.

In the I Ching and Ifá, like most divination systems, the triggering event is binary in nature. One bit for the Roman birds flying from the East or from the West; six bits for the Chinese hexagrams; eight bits for the Nigerian encounters of Odù. These events determine an interesting combinatorial setting that accompanies the rite, proving that the babalawo must have mathematical skills and, once again, showing the power of exponential growth. Sixteen special beings called Odù were trained by a god to spread wisdom through the world. The names of the Odù are coded by four-bit numbers that appear during the rite as the positions of the half-nuts thrown by the babalawo. The half-nuts can land with the concave side up (0) or the convex side up (1). The table of all Odù, ordered row-wise by their *rank*, is shown in Figure 5.1. They pay ritual visits to each other, or stay by themselves, as revealed by the combination of two groups of four nuts out of the eight nuts thrown, thus giving rise to a total of $2^8 = 256$ dispositions of their pairs, each one

---

[4]Taken from: Sergeant, R.B. *Islam*. In: Loewe M. and C. Blacker, eds. 1981. *Divination and Oracles*. George Allen & Unwin, Boston.

[5]The book of I Ching is so famous that no presentation of it is needed, except possibly recalling that the foreword to the first english translation was written by the famous psychiatrist Carl Gustav Jung who claimed to have used the book for the exploration of the inconscious. The Nigerian Ifá divination system was added in 2005 by UNESCO to its list of Masterpieces of the Oral and Intangible Heritage of Humanity.

OGBÈ	0000	ÒYÈKÚ	1111
ÌWÒRÌ	1001	ÒDÍ	0110
ÌROSÙN	0011	ÒWÓNRÌN	1100
ÒBÀRÀ	0111	ÒKÀNRÀN	1110
ÒGÚNDÁ	0001	ÒSÁ	1000
ÌKÁ	1011	ÒTÚRÚPÒN	1101
ÒTÚÁ	0100	ÌRETÈ	0010
ÒSÉ	0101	ÒFÚN	1010

FIGURE 5.1: Table of the Odù with their half-nut coding.

directing the rite that follows (the visit of one Odù to another, or the opposite visit, account for two distinct situations). So the visit of ÌWÒRÌ to ÒDÍ
is represented by the throwing 10010110, while 00000000 represents OGBÈ
(or ÈJIOGBÈ the monarch) himself. The art of the babalawo encompasses
remembering two-hundred and fifty-six different situations and knowing the
consequences of each one for the rite.

5.1 Probability theory develops

Although the most sophisticated divination systems seem to be connected
with the laws of chance, it is clear enough that the diviners never considered
things that way. And in fact, in the history of mankind, probability theory
was introduced well after divination. More or less at the time when Cardano's
book was finally published, Blaise Pascal got involved in some calculations
about playing dice. His interaction on this subject with Pierre de Fermat
gives credit to the common assertion that the two are co-fathers of modern
probability. Many facts happened later that are important for the history of
science, but not in our context until another pillar of French mathematics,
Pierre-Simon, Marquis de Laplace, published in 1812 a general treaty on the
analytical theory of probabilities, followed two years later by the amazing
little book *Essai philosophique sur les probabilités* (A Philosophical Essay on
Probabilities), where inductive reasoning is used to explain this science to a
larger public.

This book illustrates a series of basic principles of probabilities. When
discussing his sixth principle related to "bringing back each event to its cause,"
Laplace underlines that it is wrong to believe that the occurrence of a "regular
event," like the appearance of twenty heads in a row when playing head and
tail (*croix* and *pile* in French), be less probable than any other "irregular
sequence" of outcomes. And he adds the following famous sentence:

Regular combinations occur less frequently only because they are less numerous. If we look for a cause when we perceive symmetry, however, it is not because we consider a symmetric event as being less possible than others, but because, being such an event the effect of a regular cause or of chance, the first assumption is more probable than the second.

Laplace left his concept of regularity at an intuitive level. More than one century had to pass before this concept could be defined formally, that happened after the tools of algorithms and complexity had been developed. Still, as we shall see, Laplace's intuition was very acute in several senses.

Cardano's intuitive definition of probability as the ratio of the favorable events over all possible events survives in Laplace's first principle, although the other principles form a much richer body. Over the years this definition turned out to be ambiguous and gave rise to many paradoxes, particularly because it is not clear, except for obvious cases like coin tossing or casting a die, what a random choice means. The crucial advance came only in 1933 when Kolmogorov published a now famous book on the *Foundations of the Theory of Probability*, where this theory was formulated on an axiomatic basis. The probability is now the measure of an event and must comply with certain basic axioms. Although fundamental for the progress of science, we will not go further on this topic because some elementary combinatorial considerations are sufficient for our study. Instead let us go back to Laplace's sentence on "regular combinations" that contains a wealth of implications for the theory of probability.

First, it is assumed that chance *exists*, and is in fact the "cause of the generation of random combinations." However the assumption is not at all obvious and the question of its validity remains essentially open. People have different opinions. Carl Gustav Jung and other famous psychologists had no doubt on the existence of chance, that was regarded as a threat to rational behavior. Physicists assert its existence to support part of their theories, at least until other theories prove more valid. Biologists are not completely convinced. Even from a philosophical point of view it may be hard to accept the existence of chance because a random event is assumed to be independent of the past history, thus assuming that the world is well disposed to occasionally renouncing its memory.

We will escape the perils of this discussion by personifying chance as a deity of our private Olympus, as the Greeks and the Yoruba would have done. Following Laplace, for us chance exists and that's that! Among other consequences, this assumption implies that tossing an honest coin n times produces one out of the 2^n possible sequences of heads and tails with a probability of $1/2^n$. This process is called a Bernoulli trial. Each of the two binary sequences of thirty-one tosses reported in Figure 5.2 has a probability of $1/2^{31}$ (less than one over one billion) of appearing, even though they look very different because the first one consists of fifteen repetitions of the subsequence 01 followed by 0 while the second does not exhibit any apparent regularity.

0 1 0 1 0 1 0 1 0 1 0 1 0 1 0 1 0 1 0 1 0 1 0 1 0 1 0 1 0 1 0

1 0 1 0 1 0 0 0 1 1 0 0 1 0 1 1 0 1 1 0 0 1 1 0 0 0 0 0 1 1 1

FIGURE 5.2: Bernoulli trials with "heads" and "tails" coded by 0 and 1. The first sequence has a "succinct" description. What about the second?

Laplace uses the words regular and symmetric to indicate the existence of a sort of rule that a sequence obeys in its composition. For example, with the standard notation of formal languages, the first sequence of Figure 5.2 follows the rule (or, can be expressed as) $(01)^{15}0$, with the obvious meaning. Then in Laplace's terms this is a "regular combination." For the second sequence, however, we cannot easily find a formation rule and are inclined to conclude that the sequence is the result of pure chance. In fact this sequence represents "the first thirty-one digits of the binary expansion of the decimal part of π," that is the number 1415926535 expressed in binary, that is clearly not a random number. Then even in this case there is a generating rule, although not so "simple." Expressed this way the concept is very vague, but it has a precise mathematical meaning as will be made clear shortly.

Two observations, made by Laplace on a purely intuitive basis, have been proved rigorously true one and a half centuries later. The first is that "regular combinations" occur rarely because they "are less numerous." The second follows a preliminary assertion that all possible events can be generated by a random source, so that an honest coin tossing may produce the two sequences of Figure 5.2 with identical probabilities. Notwithstanding this, Laplace says, when an event can be explained by a "regular cause" we are led to believe that in fact this is what happened. We will now see how some bright mathematicians have satisfactorily answered these questions, allowing a much deeper understanding of random phenomena.

To this end, and perhaps to the surprise of Laplace in his grave, the concept of randomness has been separated from the generating source to become a property of the event itself, no matter how it came about, thereby putting to one side the philosophical issue of the existence of chance. In this new perspective we are able to affirm that *random sequences do exist* independently of their source. To get to this point, however, we must proceed a little more in our history.

5.2   Randomness as incompressibility

Aside from pure probability theory, random sequences have been thoroughly studied in the mid twentieth century in the realm of Information Theory, a branch of science devoted to studying the information content of messages and initially developed by C.E. Shannon and D.A. Huffman. Following their approach the *entropy* of a source was introduced as a measure of the *degree of uncertainty* with which the messages generated by a source may appear. Without getting into mathematical definitions, we notice that if all messages have the same probability of occurring the source has maximal entropy, while if a message occurs with probability one, i.e., with complete certainty, and all the other messages will never appear, then the entropy of the source is zero. That is consistent with the observation that no new information can be extracted from a source whose output is known a priori.

All messages are represented with the symbols of an arbitrary alphabet. They become binary sequences when representing coin tosses; or musical notes on a song score; or sentences in the Latin alphabet as in this text. So messages, mathematical definitions, and computing algorithms, are all citizens in the same world of sequences if a common alphabet is fixed, and since this alphabet is arbitrary we will refer to the binary one. On these grounds Information Theory took an unexpected direction in the 1960s with the computational approach independently adopted first by Solomonoff in the U.S.A., then by Kolmogorov in Russia and Chaitin in the U.S.A. again. The concept of *algorithmic complexity* of a sequence arose, leading to a new characterization of randomness that is no longer a consequence of the source but entirely depends on the inner nature of the sequence. As the name suggests these studies are strictly dependent on the theory of algorithms that had begun to develop in the preceeding decades. This is the new tool by which many intuitions of the past were rephrased in solid mathematical terms.[6]

The reference system is strictly mathematical and quite sophisticated. Computations are ideally performed on Turing Machines that, as we have already seen, may execute any of an infinite set of algorithms. In practical terms the machine can be any computer if no limitation is put on the memory size, and the algorithms may be coded in any programming language. A point to be remembered is that, in any family of machines, there always exists at least one that is *universal* in the sense that it can simulate the computation of any other machine that executes any one of its programs. In the real world any reasonable computer accepting programs written in a general purpose programming language as Java or C is universal provided its memory is large enough to execute any program at hand.

Sequences to work on, machines, and algorithms, are described with the characters of the same (binary) alphabet. Each sequence may be generated by any machine with an elementary algorithm that contains the message itself and a command to output it. But of course a more skilled algorithm might exist to reconstruct the sequence by using a smarter computation. We can

---

[6]The characterization of random sequences has essentially followed the three intersecting lines of *stochasticness, incompressibility,* and *typicality,* all strictly connected with the theory of algorithms. Here we refer to the second line that is mostly related to computer science. A rigorous study of randomness is hard - see the bibliographical notes for references.

apply these concepts informally to the two sequences of figure 5.2. The first one may be generated by a skilled "program" such as:

put {expand (01)^{15} followed by 0} (5.1)

while for the second we would probably opt for the naive solution:

put {1 0 1 0 1 0 0 0 1 1 0 0 1 0 1 1 0 1 1 0 0 1 1 0 0 0 0 0 1 1 1} (5.2)

where the whole sequence is explicitly shown, because computing a prefix of the decimal expansion of π is definitely more complicated.

Once the reference system has been chosen, a generating algorithm has nothing to do with the source of the sequence. Moreover different algorithms may generate the same sequence.[7] On these grounds Kolmogorov and Chaitin defined the *algorithmic complexity* of a sequence S as *the length of the shortest algorithm that generates S* (i.e., the length of the sequence representing such an algorithm). From this followed that: *a random sequence is one having complexity "substantially" equal to its length* (i.e., the length of the naive program above). Conversely: *a sequence is nonrandom if it admits a generating algorithm shorter than the sequence itself.*

Put this way, it seems too simple. In fact the informal statements above convey ideas that stand behind a complex mathematical theory. Before going into more depth, however, we note that a sequence like the first one above, possibly extended to a large number $2n+1$ of bits, is now declared non random because the generating program:

put {expand (01)^{n} followed by 0} (5.3)

is certainly shorter than the sequence itself for large enough n.[8] Laplace suggested that one such a sequence be non random because, more than likely, it was generated by a non random source. The new definition, instead, declares that the sequence is non random on the sole basis of its structure: it could have been generated by a purely random source, although with very low probability, but still it must be considered non random.

The second sequence of Figure 5.2 is too short for drawing a firm conclusion. However, if the sequence is extended to represent n binary digits of π, for n sufficiently large, a program to compute it would be definitely shorter than the sequence itself. Then even the second sequence is non random, as it may be expected by its definition.[9] But then, do random sequences exist under this definition? An affirmative answer will be given shortly through a schematic

---

[7]In fact a sequence can be generated by an infinite number of algorithms because, once one such an algorithm is found (e.g., the naive algorithm that specifies the sequence explicitly), another algorithm can be formed from the former one with the addition of a sentence that has no effect on the output, and so on.

[8]The length of this program is of order $\log n + c$, where c is a constant, because the only part that depends on n is n itself appearing as an exponent, and the binary representation of n requires $\log n$ bits.

[9]The length of this program would also be of order $\log n + c$ with c constant, because the only part that depends on n is the specification of the number of output digits required.

introduction to Kolmogorov and Chaitin theory. Readers interested only in a general picture of the field may jump to the discussion on data compression in the next section, and perhaps return here later.

Enumerate all the binary sequences $\sigma_0, \sigma_1, \sigma_2, \ldots$ in *canonical order* as explained in Chapter 4 (for clarity we repeat this ordering here):

σ_0	σ_1	σ_2	σ_3	σ_4	σ_5	σ_6	σ_7	σ_8	...	σ_{14}	σ_{15}	...
ϵ	0	1	00	01	10	11	000	001	...	111	0000	...	(5.4)

Now consider a (possibly infinite) family $S_0, S_1, S_2, \ldots$ of computing systems, and let p be a *program* for S_i. We write: $S_i(p) = \sigma$ to indicate that the computation of S_i with program p produces the sequence σ. Recall also that p is represented by a binary sequence and let $|p|, |\sigma|$ denote the length (number of bits) of the two sequences. At the basis of the theory stands the concept of algorithmic complexity of σ, first defined with respect to a computing system S_i as:

$$K_i(\sigma) = \min\{|p| : S_i(p) = \sigma\}. \tag{5.5}$$

That is the complexity of σ is the length of the shortest program that generates this sequence using S_i. What makes this definition not so interesting is its dependence on S_i. This seems inevitable at a first glance because, in general terms, a program is written for a specific machine. However, a brilliant *invariance theorem* proves that we can get rid of such dependence if we refer to a universal system S_u that simulates the functioning of all the others with a complexity that is essentially the same. Without getting into mathematical details for which the reader may refer to the bibliographical notes, the theorem states that, for any system S_i:

$$K_u(\sigma) \leq K_i(\sigma) + k, \text{ where } k \text{ is a constant.} \tag{5.6}$$

Any program p that generates σ on S_u is called a *description* of σ. By definition $K_u(\sigma)$ is the shortest description of σ and, after relation (5.6), is taken as the *algorithmic complexity* of the sequence with precision up to a constant. This value is now simply denoted as $K(\sigma)$ and often called *Kolmogorov complexity* of the sequence. It is also customary to say that this complexity represents the information contents of σ.

At this point a concept related to the Internet steps in. Having put any sequence σ in relation with a program p that generates σ, we may decide to transmit this sequence from a node A to a node B installing proper software into the two nodes so that σ is "coded" in A as p, transmitted to B in this form, and then "decoded" in B obtaining σ again. Clearly we can hope to gain something only if $|p| < |\sigma|$. In this case we talk of *data compression*. While specific compression techniques are used in practice, the concept of algorithmic complexity is crucial for understanding what is possible.

Although the expansion of π has many statistical characteristics of a random string it is obviously non random.

Recalling that σ can be always generated by a trivial program that contains σ inside and simply transfers it to the output, and noting that the length of this program is $|\sigma| + c$, where c is the constant number of bits to code the output operation (e.g., the word **put** plus the two parentheses $\{\}$ in program (5.2)), we conclude that:

$$K(\sigma) \leq |\sigma| + c. \tag{5.7}$$

Of course $K(\sigma)$ is smaller than this limit value if there is a better program to generate σ, and in particular may be much smaller than $|\sigma|$. The question is: how many sequences can be actually compressed, and by how much? Now Laplace's intuition on the scarcity of "regular combinations" will be confirmed.

It turns out there are not enough short descriptions for all sequences, no matter how long such sequences may be. In fact there are 2^n binary sequences of length n and $1 + 2 + ... + 2^{n-1} = 2^n - 1$ binary sequences of length from 1 to $n - 1$, then there is at least one sequence σ of length n without a shorter description. This σ is incompressible, but this is not the end of the story. There are $1 + 2 + ... + 2^{n-2} = 2^{n-1} - 1 = 2^n/2 - 1$ binary sequences of length at most $n - 1$, then about one half of the sequences of length n cannot be compressed to less that $n - 1$ bits. Similarly about $3/4$ of the sequences of length n cannot be compressed to less than $n - 2$ bits, and so on. Out of all possible sequences, the compressible ones are a small minority.[10]

It is now quite natural to call a sequence σ that cannot be compressed, that is $K(\sigma) \geq |\sigma|$, random. However there are subtle theoretical reasons indicating that a strictly valid definition of randomness can be given only for infinite sequences, and furthermore, pure incompressibility is too demanding a property on practical grounds because there is no interest in compressing a sequence of n bits to, say, not less than $n - 1$ bits. Therefore for a parameter $f > 0$ that may be a constant or a function of n, σ is called *f-random* if it cannot be compressed to less than $n - f$ bits, that is $K(\sigma) \geq |\sigma| - f$. The logarithmic function is a standard for f, so we loosely say:

A finite sequence σ is random
if and only if $K(\sigma) \geq |\sigma| - \lceil \log_2 |\sigma| \rceil$. $\tag{5.8}$

From what we have seen about the number of compressible sequences, the statement above is an implicit proof that *random sequences do exist*, and that they are in the vast majority among all possible sequences.

Taking these concepts intuitively, imagine that our world of sequences is the one of well formed sentences in English and the formal system at hand is a series of commands to locate different sentences within the books of a vast library. Nobody has ever thought that the sentence of fifty-three characters:

---

[10]In particular any sequence representing a program p that constitutes a minimal description of a sequence σ, i.e., $K(\sigma) = |p|$, is incompressible. Otherwise p should admit a shorter description p' that in turn would allow to reconstruct σ, against the hypothesis that p is minimal for σ.

It is an ancient Mariner
and he stoppeth one of three

was created at random, although this may have happened with the minuscule probability of $1/27^{53}$ after the draw of fifty-three characters from an alphabet of twenty-seven including the blank. In our system the sequence has to be considered non random because it can be univocally determined by a short description like $A335.h15.1, v.1\text{-}2$, that gives the library accession number to the poem "The Rime of the Ancient Mariner" by Samuel Taylor Coleridge, with the addition of "v.1-2" to indicate the first two verses. With the same reasoning the whole poem is non random, but the sequence of the first two words "It is" must be taken as being random, since it has no shorter description than the one that explicitly spells it out.

Statement (5.8) defines randomness independently of a generating cause. This is no more than an arbitrary definition, but it turns out to be very important in probability theory where, incidentally, it leads to the same conclusions as other reasonable models. In particular it has been proved in the realm of information theory that, for the length of the sequences tending to infinity, the Kolmogorov complexity of a sequence tends to the value of the source entropy. This is a nice touch for assessing the soundness of both theories. Furthermore this definition of randomness gives a mathematical sense to Laplace's assertion that we are led to believe that a regular (now: compressible) sequence σ of length n was generated by a specific rule (now: a short description p) rather than randomly. We are in the case: $K(\sigma) = |p| < n$. As 2^n sequences may be generated at random, and only $2^{|p|}$ of them can be generated by a rule of length $|p|$, it is $2^{n-|p|}$ times more likely that σ was generated by a non random cause.

The existence of random sequences is good news for many fields of applied science where such sequences are needed for simulating physical events, or for building cryptographic keys, etc. But, as in many jokes, after good news there is bad news. Mathematical logic gives a theoretical limit to the discovery of random sequences because *the problem of establishing if an arbitrary sequence is random is undecidable.* That is, if a short description is found we conclude that the sequence is nonrandom, if such a description is not found we will never know whether it does not exist or we were just unable to find it. Readers with interest in computability may enjoy a rigorous proof of this fact, among the most elegant in the field. We show here the structure of this proof, built on the so called Berry's paradox on integers of 1908.

For the sequences $\sigma_0, \sigma_1, \sigma_2....$ listed in (5.4), assume that there exists a program $R(i)$ to decide whether an arbitrary sequence σ_i is random. Take now the new program P of Figure 5.3. Although the existence of R is merely assumed, the length $|P| + |R|$ of the whole program is independent of the sequences σ_i treated at each iteration, no matter what R looks like. Since such sequences are treated for increasing length as in the listing (5.4), the condition $|\sigma_i| > |R| + |P|$ must be verified from a certain value of i on. Furthermore we

program P
  for $i = 1$ **to** ∞
    if $\{(R(i) = \text{true}) \cap (|\sigma_i| > |R| + |P|)\}$
    print(i) **and stop;**

FIGURE 5.3: A contradictory program to prove that randomness cannot be decided algorithmically.

have seen that random sequences of any length exist, so the **stop** condition on P must eventually be met on a sequence σ_i that is both random ($R(i)$ = **true**) and longer than the program that allows it to be detected ($|\sigma_i| > |R| + |P|$): a contradiction. We must conclude that program R cannot exist, i.e., is impossible to decide randomness algorithmically.

This argument rules out the possibility of deciding whether a given sequence σ is random as long as a shorter description of it is not found.[11] However the use of random sequences is important in applied science, therefore in most cases they are actually generated by computer programs. Since these programs are inevitably shorter than the sequences themselves, these are non random by definition, but are useful anyway if they pass some standard suite of statistical tests. In this case the sequences must be more correctly called *pseudo-random*.

5.3 Compressing and hashing

In 1946 Jorge Luis Borges wrote a one-paragraph story entitled *Del rigor en la ciencia* (On the Rigor of Science)[12]:

In that Empire, the Art of Cartography reached such Perfection that the map of one Province alone took up the whole of a City, and the map of the empire, the whole of a Province. In time, those Unconscionable Maps did not satisfy, and the Colleges of Cartographers set up a Map of the Empire which

---

[11] A subtle doubt may arise, namely, why not trying all possible programs shorter σ to find out whether one of them actually detects this sequence? After all these programs are finite in number so the experiment is theoretically legitimate. The impossibility of following this line lies on one of the first results of Turing, i.e., is algorithmically undecidable if an arbitrary program on an arbitrary input terminates its computation in finite time (see Chapter 4). Then we simply cannot try all possible programs shorter than σ because, as long as one of them does not stop, we cannot decide whether we have to wait longer to see whether σ will be generated, or we will have to wait forever.

[12] The story was first published in Borges' *Historia Universal de la Infamia* (A Universal History of Infamy). The present English translation is from: Borges, J.L. 1999. *Collected Fictions*. Translation A. Hurley. Penguin Books.

had the size of the Empire itself and coincided with it point by point. Less Addicted to the Study of Cartography, Succeeding Generations understood that this Widespread Map was Useless and not without Impiety they abandoned it to the Inclemencies of the Sun and of the Winters. In the deserts of the West some mangled Ruins of the Map lasted on, inhabited by animals and Beggars; in the whole Country there are no other relics of the Disciplines of Geography.

Clearly the map of the Empire was incompressible, at least in the opinion of the cartographers. In fact we have seen that most sequences cannot be compressed. How is it, then, that our computers make a continuous use of compression algorithms for saving resources such as memory space and transmission bandwidth?

Although, in the ideal world of sequences, non randomness is a rarity, the sequences we have to deal with in everyday life are in general not at all random. Computers mostly exchange text written in natural languages, or graphical and audio information. Independently of their meaning, these files adhere to grammatical structures, or to visual laws, or to musical harmony etc., so all of them exhibit inevitable regularities that can be exploited for compression. The obvious case is a run of the same pattern like a bit, or the ASCII code of a letter, or the color of a pixel etc., if the run is long enough to induce a reduction in the sequence length as in the "program" (5.1). But of course much more sophisticated algorithms are used in practice.

First consider that even incompressible files can be compressed if we do not object loosing a small amount of information. The decompressed file may differ from the original one in some minor details while the storage occupancy, or the transmission time, is drastically reduced. This practice, known as *lossy compression*, is mostly used with audio and visual data where a certain amount of detail can be ignored, and is pushed to its extreme in real time transmission, as for example in telephony over the Internet. By contrast, *lossless compression* is used for storing or transmitting written text whenever absolute fidelity is required.

For both lossless and lossy compression some standard algorithms have been created, so that the compressed files can easily be exchanged by all users on the Internet. All such algorithms are very technical and cannot be explained in simple words, therefore, we mention here only the basic concepts on which they rely. Lossless compression of text uses two main mathematical tools. One is the coding invented by David Huffman and described in Chapter 3, where each word or character of a file is represented with a certain number of bits, such that the higher the probability of the word occurring, the fewer bits are used. The second tool, proposed in 1977 by Abraham Lempel and Jacob Ziv and associated with the shorthand LZ, essentially looks for subsequences that repeat several times in the file to be compressed and substitutes such occurrences with the specification of their positions in the compressed file. A combination of Huffman code and LZ analysis is found in *gzip*, the most widely used lossless compression utility.

Lossy compression algorithms are very sophisticated. *JPEG*, the well known and most popular standard for digital pictures, takes the sequence of bits representing the points of a raster scanning of the image and applies to it a complex mathematical transform, "rounding off" some parameters corresponding to features that are less perceived by the human eye. *MP3*, the universally accepted standard for digital audio compression and musical file sharing (with consequent issues of copyright infringements), reduces the accuracy of sounds by cutting off details that are borderline for the human perception, by means of a complicated physical transform. Both lossless and lossy compression techniques are widely used in the Internet but are not part of network technology, which is what interests us in the present context. However, an important concept that we will meet again is hashing.

A *hash function* maps the elements of a very large set A onto the ones of a much smaller set B such that, on average, a large number of elements of A correspond to any given element of B. In practice all these elements are represented by binary sequences, with the sequences of A much longer than those of B. Hashing can then be seen as lossy compression, although it was developed for different purposes. The original use of hashing was for storage of data items, where A is a set of all items that may occur in a given application, and B is the set of locations where the items that actually occurred are stored. In addition hashing is extensively used today for comparison and retrieval of very large files on the Internet, where each file of A has a much shorter hash image (also called "fingerprint") in B. If the fingerprints of two files are different, the files are also different. If the fingerprints are equal, other checks must be performed. To make fingerprint equality as unlikely as possible, a good hash function must distribute the elements of A over B at random so that, with very high probability, similar sequences of A are associated to distinct elements of B.

Hash functions are also used in cryptography where they must exhibit some additional properties. In this realm a family of standard functions has been developed, called SHA for *Secure Hash Algorithm*, whose second release SHA-1 is extensively used on the Internet for fingerprinting although no cryptographic capability is required here. The algorithm for SHA-1 accepts a possibly enormous sequence as input (the upper bound is $2^{64} - 1$ bits) and produces a 160 bit output in all cases. This output is usually adopted as the fingerprint of the input file. The reduction in length is huge, but still an immense family of fingerprints is available, consisting of the 2^{160} sequences of 160 bits. The algorithm is trusted to cover this whole set almost uniformly, although no mathematical proof can be given since the output is pragmatically built by a very large number of repeated operations on the input sequence such as modular addition, exclusive OR, circular shifts, and others, that are trusted to produce pseudo-random effects. As a consequence, the probability that two files have the same fingerprint is exceedingly low, even if they are very similar. In any case the use of SHA-1 on the Internet is not limited to file retrieval as we shall see below.

5.4 Randomized algorithms

One of the reasons why random sequences are relevant for us is their role in the construction of efficient algorithms. At the outset it may seem a little strange that randomness may help in designing sequences of actions aimed at reaching precise objectives, but there are cases in which relying on unpredictable events helps a great deal. We are not talking of supernatural or psychological influences as in the divination rites, but of phenomena that can be expressed in rigorous mathematical terms.

Randomization is introduced in algorithms in different ways. Random sampling can be performed on the values of a function that cannot easily be expressed in mathematical form, so that expected values of areas, volumes, different moments, etc. can be evaluated. This approach leads to the *approximation algorithms* of numerical analysis, whose forefather can be located in the eighteenth century when George-Luis Leclerc, compte de Buffon, showed how the value of π can be estimated by throwing at random a needle on a ruled sheet of paper and counting how many times the needle crossed a line.[13] However other approaches are more interesting for us.

Random choices may help to resolve some situations where the computation proceeds too slowly, provided that the correctness of the final result is guaranteed. Algorithms of this sort are called *Las Vegas*; they lead to *certainly correct results* in a *probably short time*. Finally, and possibly more surprisingly, a different approach uses random choices at particular steps to attain *probably correct results* in a *certainly short time*. We are now allowed to abandon the certainty of correctness only if the probability of failure is negligible, an adjective that will be expressed in precise mathematical terms. This is the family of *Monte Carlo* algorithms. No concept of approximation applies here because if the result is wrong it may be completely wrong, like returning binary result as 0 when it should be 1.

To explain how these concepts get into the functioning of networks let us start from a fundamental structure of parallel processing. Figure 5.4 shows a *hypercube*, or simply *cube*, in four dimensions, with $2^4 = 16$ nodes and four arcs departing from each node. The binary names associated to the nodes will become clear shortly. A hypercube in h dimensions has 2^h nodes and h arcs per node and can be seen as the union of two hypercubes in $h - 1$ dimensions: in our example we have two three-dimensional cubes with the familiar shape of a cube, placed at the left and at the right of the picture. Each one of

---

[13]This surprising result can be actually proved quite simply. Assume a needle of length l thrown on a ruled plane with parallel lines at a distance t from one another, with $t \geq l$. It can be proved using elementary calculus that the probability P that the needle crosses one of the lines is $P = 2l/(t\pi)$, then $\pi = 2l/(tP)$. The experiment goes like this: throw the needle n times, with a large n; count the number h of crossings; and take $P \approx h/n$. This leads to the estimate: $\pi \approx 2ln/(th)$. An excellent source for understanding this computation is Wikipedia at: http://en.wikipedia.org/wiki/Buffon's_needle

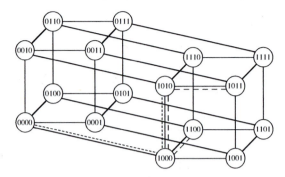

FIGURE 5.4: A hypercube in four dimensions. The two standard routes, one from node 0000 to 1010 (dotted) and the other from 1100 to 1011 (dashed), interfere in node 1000 and in arc (1000–1010).

them has $2^3 = 8$ nodes and three arcs for each node, plus an additional arc connecting each node with one in the same position of the other cube to make the structure grow into the forth dimension. The first bit of each node name identifies the two cubes, being 0 in one of them and 1 in the other. Each three-dimensional cube is the union of two two-dimensional cubes (squares) with $2^2 = 4$ nodes in the front and in the back of the picture, identified by the value of the second bit of the node names. These cubes are the union of two one-dimensional cubes (segments) identified by the third bit, that are finally decomposed in zero dimensions, i.e., into single nodes identified by the last bit.

Once the structure is known we can let it grow to arbitrarily high dimensions. A hypercube in five dimensions is the union of two hypercubes like the one in the figure, with the nodes pair-wise connected by a new arc. The new structure has $2^5 = 32$ nodes and five arcs per node, and the node names acquire a fifth bit to the left. The rule is simple, but the graphical representation becomes quite complicated. What is of particular importance is the neighborhood of each node. In a hypercube in h dimensions, each node has h neighbors; hence this number grows with the size of the network. In practical terms this growth cannot be accepted over certain values because the number of lines connected to a node affects the size of the node itself, no matter what the node stands for. Still for $n = 2^h$ each node has $h = \log_2 n$ neighbors, that is, this number grows very slowly with n.

The hypercube construction rule implies that any two neighbors have names that differ by exactly one bit. Therefore if two nodes have names differing by d bits they can be connected through a path of d arcs which traverses a sequence of nodes whose names differ by one bit at each step. In the hypercube of Figure 5.4 a message can be sent from node 0001 to node 1111 along various paths of three arcs, for example the one that goes through the nodes 0001-1001-1101-1111. In fact this is a *standard* route where the nodes are chosen at each step by flipping the bit names from left to right wherever required. If the name of the destination is included into the message, the routing is decided by the nodes encountered along the path. At each step the node currently reached flips a bit name to decide the neighbor to which the message has to be sent. As a conclusion a path connecting any pair of nodes has length at most $h = \log_2 n$ because two node names differ at most by h bits.

In a hypercube, then, all paths are "short," and this is an important characteristic of any network. However, message delivery may be slowed down by traffic jams, as queues of messages may pile up in particular nodes and arcs even though all of them have different origins and destinations. The crucial problem of routing will be discussed again in the subsequent chapters. Here we confine our analysis to exchanging messages in a hypercube, with the aim of understanding how randomization may help.

Assume that two messages are sent at about the same time, one from node 0000 to 1010, the other from node 1100 to 1011. The standard routes:

0000 1000 **1010**

1100 1000 1010 **1011**

are applied in Figure 5.4. They interfere in node 1000 and in arc (1000-1010). Obviously such an interference could be avoided by adopting a different routing strategy, but other interferences would occur for different messages.

A mathematical study on the formation of queues of messages under different routing rules is quite complex. We recall informally a major result that was first proved for the hypercube in h dimensions and then extended to other regular network topologies. These studies date back to the time when the first parallel computers were built, with the hypercube being the major interconnection structure of the CPUs for computing some important functions efficiently. Pathological cases were known in which the queues grow to a length exponential in h even though all the messages have different destinations, but a probabilistic argument showed that, if the destinations are chosen at random, the arising queues are no longer than h under the standard routing rule. This relieved most concerns, as that case was believed to absorb all the situations that practically occur. But then an unexpected difficulty showed up.

The new threat closely recalls the formation of vehicular queues in some urban traffic conditions, where many drivers follow some common paths even though they eventually go to different destinations. Drivers tend to cluster on preferred roads by force of habit, or common reasoning, yet they curse their peers who have selected the same roads at the same time. Unexpectedly, parallel processing tends to induce the same behavior. As we shall see when discussing Web search engines that employ huge clusters of computers, all working together, the designers of parallel algorithms have a deep-seated propensity to let all the processors in the game do similar operations at the same time (that is fine), but also exchange messages along the same routes (that is bad). So these algorithms tend to form much longer traffic queues than the ones arising with the same messages traveling along random routes.

The initial study that had a major impact on the routing strategies is due to Leslie Valiant, a bright computer scientist of Harvard University, though the ideas had been ripening in the scientific community for some time. If a message has to be sent from node A to B, send it first to another node C chosen at random and then proceed from C to B, using a standard rule in both phases. For the hypercube, and then for several other networks, it was proved that the queues thus arising have a length at most h with very high probability, where the mathematical meaning of the latter sentence will become clear in the following. The tendency of algorithm designers (but, we might say, the bias of the problems themselves) to create traffic jams is then outmaneuvered by a merry anarchy of random paths.[14]

The routing strategy just described is an example of Las Vegas processing, where a sound result is always attained in a probably short time. The alternative is to trade off correctness for efficiency without taking a leap into the unknown. In a Monte Carlo algorithm we are allowed to give up correctness only if the probability of getting a wrong result can be made smaller than an arbitrary value that typically depends on one or more parameters of the computation. Choosing a smaller probability value v must imply an only modest increase of the running time t, or of some other cost. In a good algorithm v decreases exponentially with increasing t. Such an approach may irk the purists, but becomes acceptable if the probability of failing this way is inferior to the one of getting a wrong result for any other reason like a hardware crash, an incorrect reading of the output, or a sudden attack of madness in the algorithm designer.

To make this approach clear, we start with forensic courts, where mathematics has never been used for landing people in jail or sentencing them to the death penalty. Only after the introduction of scientific proofs, such as those based on DNA analysis, has the concept of probability even being raised in judiciary debates, and only then in an unassertive and imprecise manner. In particular, conditional probability, that is the chance that an event actually occurred if another event occurred too, has very rarely been considered.[15] Our proposal is far less ambitious and takes its steps from a literary jewel.

Judges often overlook the role that chance may have in the courtrooms, with the exception of Judge Bridlegoose who, according to Franois Rabelais, duly listened to the witnesses of each case but then gave his sentence by casting

---

[14]Extending this strategy for avoiding vehicular jams in a city is an exciting possibility. Each driver should have the guts to take a direction at random, say, at each of the first h crossings encountered if the city has 2^h crossings. We cannot swear that the system will actually work because cities are not hypercubes and drivers are not computers.

[15]An interesting account of the importance of conditional probability in criminal trials is given in the article: Angela Saini, Probably guilty: bad mathematics means rough justice, *New Scientist* magazine issue 2731, London, October 2009.

dice.[16] When asked by the President of the High Court why he was behaving so unusually, Bridlegoose answered that he was doing nothing different from what all the other judges were doing:

> *But when you have done all these fine things, quoth Trinquamelle, how do you my friend, award your decrees, and pronounce judgement?*
>
> *Even as your other worships, answered Bridlegoose; for I give out sentence in his favour unto whom hath befallen the best chance by dice, judiciary, tribunian, pretorail what comes first.*

Without necessarily being supporters of Bridlegoose, let us see how a probabilistic evaluation of different testimonies may influence a trial. With this we are not pretending to teach the judges their business, or to influence the juries, but we want to address ordinary people who generally know judicial practices from crime fiction. Is a defendant really guilty beyond any reasonable doubt? In Perry Mason's TV series a lady may end up crying and confessing to be the murderer of her husband because he was so cruel, etc. But the story could go on ten minutes more, only to see her daughter confessing in tears that in fact she is the murderer, and her mother was just trying to protect her. As this chain of hypotheses could go on without an end, let us ask ourselves at what point it is proper to conclude a trial and give a sentence with the aid of a Monte Carlo algorithm. Obviously we are making a joke, but is useful to understand how these algorithms work.

To be liberal (but also to simplify the computation) we assume that one testimony in favor of the defendant is sufficient to acquit, while a testimony against has to be taken with caution. The lower the probability that the witness is deceptive, the stricter the proof of guilt. The value of these probabilities should be assigned by the judge, who is supposed to admit only testimonies that are *independent* from one another. If n independent witnesses of guilt have been listened to, and $p_1, p_2 ..., p_n$ are the probabilities that their sworn statements are false, the probability that the defendant is unjustly condemned is given by the product $P = p_1 \times p_2 \times ... \times p_n$ that is the probability that n independent events occur. Recall that all p_i are less than one, so their product decreases for increasing n.

The responsibility of the judge is now apparent. No sentence is absolutely certain, however, sufficient testimonies must be accumulated until the probability of error becomes smaller than the probabilitiy that other accidents render the sentence ineffective, for example a power blackout that interrupts electrocution. If the testimonies are independent (otherwise we cannot multiply the corresponding probabilities) a small number of them are sufficient to attain a very small probability of error. In particular, if $p_i = 1/2$ for all i we have $P = 1/2^n$, a value that decreases exponentially with n (i.e., with the "running time") as required in any good randomized algorithm. We do not

---

[16]F. Rabelais, *Gargantua and Pantagruel*, III, 39.

pretend that our proposal on how to conduct a trial is perfect, but after all we do better than Bridlegoose, who only gave a correct sentence every other case.

5.5   Example: file sharing on the Internet

Randomized algorithms have been developed by computer scientists for solving their own problems that, although less dramatic than sentencing a defendant to death, are nevertheless worthy of study. As an example we pick the problem of file sharing on the Internet as one of the most interesting, simplifying its mathematical analysis as much as possible. The overall structure is of a group of users with equal resources and rights (*peers*) connected to a network where they download files of interest directly from each other. Today this type of communication is mostly devoted to obtaining music files in MP3 format, and hashing is a standard technique to allocate and retrieve these files.

The most popular peer-to-peer file sharing systems are decentralized, that is they are fully distributed among the users. There is no central server hosting a complete set of addressing tables to route a message between users, rather each user stores a limited amount of routing information locally.[17] Furthermore all users execute the same protocols for locating and downloading files. For maximum efficiency the files must be stored at the user sites as evenly as possible, coping with the non trivial problems of relocating the files of a user who leaves the system, or assigning files to a new user. In fact standard hash functions are very efficient for storing and retrieving items in a static environment, but if the storage buckets change dynamically, nearly all items must be relocated at each change. The problem then is finding a scheme where most of the items remain in their buckets after the insertion or the removal of one bucket from the system. To this end a randomized allocation algorithm called *consistent hashing* works nicely, but the results are possibly incorrect with a very low probability. The algorithm was originally designed for distributing files in a dynamic family of Web servers, and is currently used in a wealth of distributed applications to cope with system changes and failures.

Roughly speaking the problem is to evenly allocate M items (files) into N buckets (peer users) connected in a network such that a change in the set of buckets requires only M/N items to be relocated on average (note that M/N is the expected number of items in one bucket). Furthermore this must

---

[17] According to the way it is conducted, file sharing may raise legal problems that it is not our role to investigate. The decentralized structure was originally chosen to circumvent these problems.

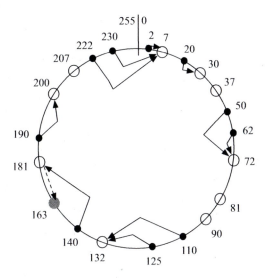

FIGURE 5.5: The circle of consistent hashing for $C = 2^8 = 256$. Buckets and items are represented with white circles and black dots, respectively. Solid arrows indicate the destination of items. The grey circle 163 indicates a new bucket receiving items from its successor 181 (dashed arrow).

be associated with an efficient method for file retrieval. Consistent hashing meets these requirements in an elegant way.

Having chosen an integer C to represent the maximum number of buckets allowed to participate in the game, the whole system can be represented on a circle that hosts the interval $[0,C\text{-}1]$, wrapped clockwise. The buckets and the items are randomly mapped to the integers in $[0,C\text{-}1]$, then buckets and items are represented as points on the circle. An elementary example is given in Figure 5.5 with only $N = 10$ buckets and $M = 10$ items out of a maximum of 256 each. Each item is allocated into the closest bucket encountered clockwise around the circle. A new bucket entering the system is mapped randomly on the circle and receives appropriate items from the successor. A departing bucket sends all its items to the successor.

Since the mapping of buckets and items to the circle is random, the expected number of items per bucket is M/N. Therefore roughly $M/2N$ items are inserted into a new bucket, and M/N items are released by a departing bucket. Since these item relocations take place between two buckets only, while a participation of more buckets would ensure a more balanced distribution, consistent hashing makes use of a more complex strategy where buckets are replicated randomly in several copies along the circle. With this and other details that we skip here the whole algorithm attains some important results that we summarize below, taking into account the following.

As the number of buckets is continuously changing, N is the value relative to each *view* of the system, i.e., to the set of buckets existing at each moment. We assume $N \geq C/k$ for some constant k, i.e., each view must contain at least a certain fraction of all the possible buckets. For a given item i, the *spread* $\sigma(i)$ is the number of different buckets to which i is mapped over all the views, and the spread σ of the distribution is the maximum spread among the items. For a given bucket b, the *load* $\lambda(b)$ is the number of different items assigned to b over all the views, and the load λ of the distribution is the maximum load among the buckets. Note that spread and load are strong indicators of the quality of the distribution, and should be kept as low as possible. For consistent hashing we have:

- the protocol is *fast*, i.e., the expected time to map an item to a bucket is of order $O(1)$ (constant), and to add or delete a bucket is of order $O(\log C)$;

- the distribution is *balanced*, i.e., the probability that any given item i goes to any given bucket b is $1/N$;

- σ and λ are of order $O(k \log C)$ with probability $\geq 1 - 1/C$.

The values of σ and λ depend on the size of the views through the value of k (large views imply small k); and depend modestly on the maximum number of buckets, as they grow with $\log C$. In fact C is regarded as a free parameter. The probability $1/C$ that σ or λ exceeds the expected order of magnitude decreases exponentially as the term $\log C$ increases. For example increasing C from 2^n to 2^{n+k} the probability of error is divided by 2^k while $\log_2 C$ grows only to $n + k$.

Let us now see how consistent hashing can be used for file retrieval in a peer-to-peer system. We assume that a bucket is identified by the Internet address (or IP address) of the corresponding user and an item is identified by the name of the corresponding file.[18] A standard choice is the adoption of the hash function SHA-1 to map users and files to the circle, so up to $C = 2^{160}$ points are usable. For simplicity we take $C = 2^8$ in the examples, together with a random mapping R to the interval $[0,255]$. So a user with IP address $A(u)$ is mapped to $R(A(u)) = 132$; and the MP3 file of "Isla Bonita" is mapped to $R(IslaBonita) = 110$ (from now on users and files will be denoted by their positions). Again files are assigned to the closest user clockwise, so in Figure 5.5 file 110 would be stored at user 132. If a user joins or leaves

---

[18]In particular we refer to a major file retrieval service called *Chord*, described here in its main lines. For a complete description of Chord see the bibliographical notes. Note that unlike in consistent hashing, users are now mapped only once on the circle. Recall that an IP address is a number associated with each computer directly connected to the network; see subsequent chapters.

the system, files are relocated as explained before. Let us see now how lookup is accomplished. For a user u looking for file f the general strategy is the following:

1. u sends the request to p that is the closest predecessor of f;

2. p passes the request to its successor s which contains f;

3. s sends f to u whose IP address is contained in the request.

Referring to Figure 5.5, user $u = 200$ looking for file $f = 110$ must send the request to $p = 90$ (predecessor of 110), and 90 passes the request to $s = 132$ which contains the file. 132 then sends f to 200. The problem is, how can each user know the positions of the other users on the circle in order to find predecessors and successors? In the example, how does 200 find 90, and how does 90 find 132?

Since the protocol is intended for a distributed system, the users cannot interrogate a central server to discover the positions of their peers, so this information must be stored by the users themselves. For $N \leq 2^m$ a good tradeoff between storage space and search speed is obtained by assigning to each user a set of m positions (*fingers*) of other users, together with their IP addresses.[19] For $i = 0$ to m-1, the finger $z(i)$ of user u is the position of the user closest to $w(i) = u + 2^i$ clockwise, where the addition is taken modulo C. The fingers for users 72 and 200 are shown in Figure 5.6, e.g., for user 72 we have:

$w(0) = (72 + 2^0) \bmod 256 = 73$, hence $z(0) = 81$
  (note that $z(0)$ is always the successor of u);

$w(4) = (72 + 2^4) \bmod 256 = 86$, hence $z(4) = 90$;

$w(6) = (72 + 2^6) \bmod 256 = 138$, hence $z(6) = 181$; etc.

For user 200 we have:

$w(1) = (200 + 2^1) \bmod 256 = 202$, hence $z(1) = 207$;

$w(6) = (200 + 2^6) \bmod 256 = 8$, hence $z(6) = 30$;

$w(7) = (200 + 2^7) \bmod 256 = 72$, hence $z(7) = 72$; etc.

To look for a file f, user u sends a request to the user x that, according to its fingers, is the closest predecessor of f, and asks x to look for f. As u has knowledge of only $m \sim \log_2 N$ of its peers, more than likely x is not the real closest predecessor of f. The request then goes on from x in the circle with the same strategy, and proceeds through other users until f is found and sent back to u. The previous example may be followed on Figure 5.6. To retrieve file 110, user 200 searches for the predecessor of 110 among its fingers finding 72, and sends the request to the IP addresses of 72 that is stored with

---

[19]Recall that N changes continuously so we can only fix an upper bound 2^m for it.

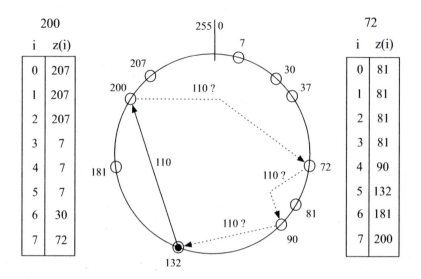

FIGURE 5.6: Fingers z(i) for users 200 and 72. User 200 retrieves Isla Bonita (file 110 stored at user 132) using these fingers.

the fingers. 72 in turn looks for the predecessor of 110 among its fingers and sends the request to 90. Since the successor of 90 is 132 > 110 (i.e., 90 is the real closest predecessor of 110) the request is sent to 132 that has the file and sends it back to 200. Assuming that user 200 is the only one who likes nostalgia favorites of the 1940s and looks for an MP3 version of "Mona Lisa" with hash $R(MonaLisa) = 128$, lookup would follow the same steps up to user 132 (successor of 128), that sends back a negative answer.

The reader may notice how this algorithm echoes that of binary search explained in Chapter 4. Due to the balanced distribution of consistent hashing, the stretch of the circle on which f resides is at least halved at each iteration with high probability. So the request makes at most m hops in the network and the expected time for one lookup is $O(\log N)$.[20] Finally it is worth noting that a limited number of user changes do not affect the system too much, so that finger tables may be kept for a long time before being updated. In addition the system is resistant to disconnection by random changes because each user has $O(\log N)$ connections to other users.

---

[20]The search for a predecessor in a finger table can be done in additional $O(\log m) = O(\log \log N)$ time using binary search, which leaves the overall lookup time unchanged. Of course our presentation of the protocol has been oversimplified. In particular insertions and deletions of users require $O(\log^2 N)$ time.

5.6   Randomness and humans (instead of computers)

We opened this chapter by talking about gambling and divination as the most common human activities associated with randomness. Now that we know much more about random phenomena we may reconsider our perception of them.

First, in the history of man divination has been more important than one may suspect. The ancient populations of Eurasia, and later of North America, held divinatory rites to direct hunting expeditions along paths where game could be found. Such rites had a dramatic importance, for the repeated failure of hunts might have lead to the extinction of a tribe. Many peoples used scapulimancy that consisted of the burning of an animal shoulder bone and the interpretation of the resulting cracks and spots to choose the direction to be taken. In the words of a scholar on the Naskapy population of Labrador:[21]

If it may be assumed that there is some interplay between the animals [the hunters] seek and the hunts they undertake [....] then there may be a marked advantage in avoiding a fixed pattern in hunting. Unwitting regularities in behavior provide a basis for anticipatory response. For instance, animals that are "overhunted" are likely to become sensitized to human beings and hence quick to take evasion actions. Because the occurrence of cracks and spots in the shoulder blade and the distribution of game are in all likelihood independent events, i.e., the former is unrelated to the outcome of past hunts, it would seem that a certain amount of irregularity would be introduced into the Naskapi hunting patterns by this mechanism.

So, in choosing a random path, the hunters may be guided by the wisdom of the shaman by more than merely his connection with the supernatural world. To benefit from randomness, computer scientists look for a certainly correct result in a probably short time (Las Vegas); or for a probably correct result in a certainly short time (Monte Carlo); or aim at the more sophisticated objective of getting a probably correct result in a probably short time, as in the peer-to-peer file sharing protocol seen above. The Labrador hunting strategy followed this last approach long before computers were invented. The benefits of randomization will be fully exploited once the success of a computer algorithm reaches the dramatic importance of a hunting expedition.

Another important consideration comes from a more natural (i.e., less mathematical) perception of "information." While considering randomness, we have come to the implicit conclusion that the information carried by a sequence can be found in any of its generating algorithms, so the length of

---

[21]O.K. Moore: "Divination: a new perspective." American Anthropologist, 59, 1957, pp. 69-74. This essay was brought to the attention of computer scientists by Jeffrey Shallit, see bibliographic notes.

the shortest algorithmic description of a message can be taken as a measure of the information contents of the message itself. Adopting this point of view, a non random message can be substituted by a shorter one at the sole cost of coding/decoding. If compression can be pushed as far as obtaining the shortest possible description, then the characters (bits) of this description are all essential, otherwise an even shorter description would exist. For the sake of practicality, if a message is incompressible it can be always identified by a shorter hash image at the cost of loosing part of the information carried by the original. This theory is strictly mathematical and has been a strong basis for several impressive achievements in data compression and telecommunications. Subtler questions arise if we regard messages from a different perspective.

Consider how nature has chosen to code things before information theory was introduced by the humans. An immediate example concerns the way our own bodies are coded genetically. As explained in Chapter 3, the DNA code is inherently redundant because twenty amino acids are represented with three characters of a quaternary alphabet (the four "bases") that makes sixty-four possible combinations. But at the level of protein coding, i.e., for those DNA stretches known as genes, nature has attained a remarkable efficiency. In the simplest organisms known as prokaryotes, e.g., in bacteria, DNA is almost entirely composed of genes; these genes entirely code for proteins and their sequences have proved to be practically incompressible. Therefore these genes have the characteristics of random sequences, a conclusion that seems hard to accept as they are also a fundamental part of the description of a fully functional organism. For the same reason, however, genes have maximum information contents. In a sense these sequences seem to have been "optimized" during the organism's evolution, excluding any redundancy.

In more complex organisms known as eukaryotes, e.g., in human beings, the situation is more complicated because genes are composed of alternating stretches called *esons* and *introns* that respectively code for proteins and contain different (and still partially unknown) information. Again the coding portion cannot be compressed, while the introns contain many repetitions of same subsequences adjacent to each other (*tandem repeats*) in specific locations (*loci*) that make introns LZ compressible. More interestingly, the number of repeats in the different loci varies between individuals, making the count of repeats a standard tool in forensic cases. Here the concept of fingerprinting comes in again. The current knowledge of the human genome has allowed the identification of some significant loci where repeats occur in numbers independent from each other, so that a genetic profile of an individual can be created by counting all such repeats. As for SHA-1, then, a short fingerprint is thereby associated to a huge DNA sequence. Albeit short, this fingerprint is sufficient to identify an individual with very high probability, and is also used to assess the identity of kin of two individuals because the number of repeats tends to be conserved among relatives.

Taking now a completely different point of view, consider how a sequence of any kind is perceived by humans. Unlike in the realm of mass media where pluralism is advocated, a mathematical definition of information must be unambiguous. In cognitive sciences, however, it may be difficult to maintain this axiom. When a message is received it adds to a patrimony of knowledge inside each one of us so the information conveyed by the message depends on the knowledge already present inside the receiver.

A similar concept is indeed contained in Shannon's theory where a source may be "non stationary," implying that the probability of a character occurring in a sequence may depend on the characters that the source has emitted up to that point, as happens for example in any natural language. However the patrimony of knowledge is acquired only from the (possibly random) source itself, whose existence has to be postulated. How does our culture, or the surrounding environment, or even our genetic heredity, influence the amount of information that a message adds to our knowledge? Do we necessarily perceive a (mathematically) non random message as such? And, most important of all:

A random message mathematically conveys the maximal possible information, but what our perception of it would be? We would likely discard such a message as meaningless, and nothing of it will remain inside ourselves.

Straightening out all these questions formally is a challenging open problem.

Bibliographic notes

Lucretius' poem *De Rerum Natura* (On the Nature of Things) can be found in several English editions, and can be also downloaded from the Web. The poem is quite long and not easy to read; however, at least the excerpt on the random agitation of dust is worth a read and should be mentioned in any class on physics. It appears in Book II, verses 112 on.

The tumultuous life and the multiform interests of Gerolamo Cardano are well described in the essay: Ore, O. 1053. *Cardano, the Gambling Scholar.* Princeton University Press.

Laplace, P.S. 1819. *A Philosophical Essay on Probabilities.* Dover, New York. This early translation of Laplace's book appeared shortly after the publication of the French original edition by the famous publishing house Gauthier-Villars.

The axiomatic theory of probability appeared for the first time in 1933 in a famous book of Andrej N. Kolmogorov and can be read in English translation in: Kolmogorov, A.N. 1956. *Foundations of the Theory of Probability.* Chelsea Publishing Co., New York. An introductory treatment of this theory can be found today in any college book on probability.

The study of random sequences is rather challenging. A very good and reasonably understandable introduction is by S.B. Volchan, What is a Random Sequence? *The American Mathematical Monthly*, 109, 2002. A must for studying Kolmogorov complexity in depth is the book: Li, M. and P. Vitni. 2008. *An*

Introduction to Kolmogorov Complexity and its Applications. Springer, New York. For data compression the reader may refer to the manual: Salomon, D. 2007. *Data Compression: The Complete Reference.* Springer, London.

An excellent source of information on randomized algorithms is the book of Motwani, R. and P. Raghavan. 1995. *Randomized Algorithms.* Cambridge University Press. The example of game hunting is suggested in the witty paper: J. Shallit, Randomized Algorithms in "Primitive" Cultures, or What is the Oracle Complexity of a Dead Chicken?, ACM *SIGACT News* 23 (4) pp. 77-80, 1992. SIGACT is the Special Interest Group on Algorithms and Computation Theory of ACM, the Association for Computing Machinery.

The series of novels on the life of the two giants Gargantua and Pantagruel is a French literature classic of the sixteenth century. The edition from which have taken the excerpt is: Rabelais, F. 1952. *Gargantua and Pantagruel.* Translated by Sir T. Urquhart and P. Motteux. William Benton, Chicago.

Consistent hashing and the associated peer-to-peer lookup protocols have been developed by a very strong research group at MIT. The mathematical analysis involved is not simple. Two major references are: Karger, D. et al., Consistent hashing and random trees: distributed caching protocols for relieving hot spots on the world wide web, *Proceedings of the 29th ACM Symposium on Theory of Computing,* ACM Press 1997, 654-663. Stoica, I. et al., Chord: a scalable peer-to-peer lookup service for internet applications, *Proceedings of the 2001 Conference on Applications, Technologies, Architectures, and Protocols for Computer Communication,* ACM Press 2001, 149-160. It is worth noting that Daniel "Danny" Lewin, one of the authors of the first paper, was killed aboard the AA11 flight during the 9/11 attacks, apparently while he was fighting the hijackers. A square in Cambridge, MA, has since been renamed in his honor.

Finally, several popular science books have been written on the theory of randomness. For example: Mlodinow, L. 2008. *The Drunkards Walk: How Randomness Rules our Lives.* Pantheon Books, New York, makes very pleasant reading.

Chapter 6

Networks and graphs

How networks can grow according to different rules: but in the end the rich get richer.[1]

The word network covers a wealth of different meanings, from subterranean fungal interconnections studied in biology to human interactions studied by social scientists. Our main interests range from the physical structure of computer networks to the logical structure of the Web, but all networks share some essential concepts and properties, and graph theory is the standard mathematical tool for studying them all. So we start by looking at graphs in some detail.

We have already seen that a graph is a mathematical object consisting of a set of nodes and a set of arcs connecting pairs of nodes together. The arcs (and consequently the graph) can be directed or not. Graphically, nodes are represented with dots and arcs are represented either with arrows if the graph is directed or with line segments if the graph is undirected. Figure 6.1(a) shows an undirected graph. Since in computer memory the story varies, we shall use whichever is the most convenient representation.

The main parameters of a graph are the number of nodes n and the number of arcs m. There cannot be more arcs than there are pairs of nodes. Furthermore, if the graph is *strongly connected*, that is if every node can be reached from any other node, then the arcs cannot be too few. In this case we have: $n - 1 \leq m \leq n(n-1)/2$ in an undirected graph (if the lower bound $m = n - 1$ is met the graph is a tree, see Chapter 3); and $2(n - 1) \leq m \leq n(n - 1)$ in a directed graph. In any case, the number of arcs spans from a linear function of n (*sparse* graph) to a quadratic function of n (*dense* graph).

Note that dense graphs do not really arise when modeling real phenomena of very large size. If the number of nodes keeps increasing, it is not the case that the number of connections to a node may also increase beyond a given bound, whatever these connections may represent. In particular, if the structure of the

---

[1] This chapter is the most mathematical in nature and has been included to allow a deeper understanding of the following chapters. Readers with a serious allergy to mathematics may jump to the next chapter directly. It goes without saying, however, that they will miss something interesting.

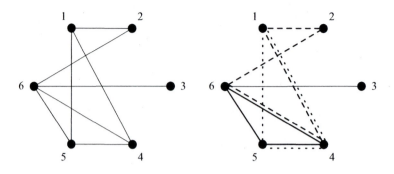

FIGURE 6.1: (a) an undirected graph with $n = 6$ nodes and $m = 8$ arcs. (b) A choice of $c = 3$ independent cycles (solid, dotted, and dashed).

nodes must be decided once and for all, the number of connections incident to each node cannot exceed a given constant k, and the total number of arcs in the corresponding graph is bounded above by $k \cdot n$. In practice very big graphs must be sparse.

For a strongly connected undirected graph the maximal number c of *independent cycles* is also relevant in the study of networks. A cycle is a closed path of consecutive arcs traversing distinct nodes, e.g., nodes 6-4-5-6 in Figure 6.1; and the cycles of a set are independent if each one of them includes at least one arc that is not contained in any of the others. For example, the three cycles shown in Figure 6.1(b) constitute a maximal set of independent cycles for that graph, although other maximal sets exist (e.g., cycle 6-4-5-6 could be replaced by 1-2-6-5-1). Essentially c is the number of edges that must be cut to ensure that no cycle remains.

The value of c is related to the values of n and m by the famous relation:

$$c = m - n + 1. \tag{6.1}$$

For the graph of Figure 6.1 we have $3 = 8 - 6 + 1$. Relation (6.1) can be proved immediately noting that, after the elimination of c arcs to cut all the cycles, the graph is reduced to a tree T with $m' = m - c$ arcs and the n original nodes. Since T is a tree we have $m - c = n - 1$ and the relation follows.

While cycles are special paths where origin and destination coincide, other paths are obviously of interest. In a directed or undirected graph the *distance* between two arbitrary nodes x, y is defined as the length l_{xy} of the *shortest path* from x to y, where the length is the number of arcs in the path. When graphs are used to represent networks the value of l_{xy} may be obviously relevant. We may also be interested in its mean value $\bar{l}$ over all pairs of nodes, i.e.:

$$\bar{l} = \frac{1}{n(n-1)} \sum_{xy} l_{xy}. \tag{6.2}$$

If there is no path from x to y we must put $l_{xy} = \infty$. So relation (6.2) has a meaning only for a strongly connected graph, directed or not. We will encounter parameter $\bar{l}$ often in the following, in particular in the context of *small worlds*, a connection structure that appears in many networks modelling real life phenomena or built as mathematical abstractions.

Another important parameter is the node degree as introduced in Chapter 1 in connection with the graph of Königsberg. In an undirected graph the degree $d(x)$ of a node x is the number of arcs incident to it. Looking again at the map of that city (Figure 1.1) we immediately recognize that Knephof island A is probably the most important district in town because it is connected to the other districts by the greatest number of bridges. That is, node A has the highest degree in the associated graph, and in fact we have seen that a random walk would end in A with the highest probability. If the graph is strongly connected all nodes have degree at least one and a node of degree one cannot belong to any cycle (see node 3 in Figure 6.1). Since each arc is incident to two nodes, the mean value $\bar{d}$ of node degree in the whole graph is $\bar{d} = 2m/n$.

In a directed graph, however, we must distinguish between the arcs entering and leaving a node x, whose numbers are respectively the *in-degree* $d_{in}(x)$ and the *out-degree* $d_{out}(x)$ of that node. Figure 6.2 shows a directed graph G where node 1 has $d_{in}(1) = 1$ and $d_{out}(1) = 3$, while node 4 has $d_{in}(4) = 3$ and $d_{out}(4) = 0$. If the graph is strongly connected all nodes must have nonzero in and out degrees, so graph G of the figure is not strongly connected (in fact no node can be reached from node 4). Since each arc is both leaving (from a node) and arriving (in another node), the mean values of the in and out degrees are the same: $\bar{d}_{in} = \bar{d}_{out} = m/n$.

In the following we shall see that the distribution of values of the node degrees is a fundamental function for understanding the structure of the graph.

6.1 The adjacency matrix and its powers

For studying a network in the form of a graph G on n nodes, the most convenient computer representation is an *adjacency matrix* A of size $n \times n$. Let the nodes be numbered from 1 to n. The rows and the columns of A are put into correspondence with the nodes of G. Cell $A[i, j]$ in row i and column j contains 1 if an arc of G connects node i to node j, contains 0 otherwise. A directed graph G and its matrix A are shown in Figure 6.2. For example there is an arc from 2 to 6 then we have $A[2, 6] = 1$; there is no arc from 6 to 2 then $A[6, 2] = 0$.

A first remark is that the in-degree of a node x is given by the number of 1s contained in column x of the adjacency matrix. In our example there are

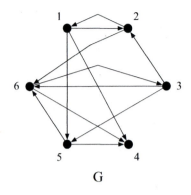

G

	1	2	3	4	5	6
1	0	1	0	1	1	0
2	1	0	0	0	0	1
3	0	1	0	0	1	1
4	0	0	0	0	0	0
5	0	0	0	1	0	1
6	0	0	1	1	0	0

A

	1	2	3	4	5	6
1	1	0	0	1	0	2
2	0	1	1	2	1	0
3	1	0	1	2	0	2
4	0	0	0	0	0	0
5	0	0	1	1	0	0
6	0	1	0	0	1	1

A^2

FIGURE 6.2: A directed graph G, its adjacency matrix A, and the squared matrix A^2 that reports the number of all paths of length two.

three 1s in the forth column then $d_{in}(4) = 3$. Similarly the out-degree of x is given by the number of 1s in row x, for example $d_{out}(4) = 0$.

An undirected graph can be seen as a special case of a directed graph if every arc of the former is replaced by two arcs of the latter connecting the same nodes but with opposite orientations. Then, unlike in Figure 6.2, the adjacency matrix of an undirected graph is symmetrical around the top-left to bottom-right diagonal, i.e., $A[i, j] = A[j, i]$ for any pair i, j. Let us play a little with these matrices before going on.

The first concept that we will need is the one of *inner-product* (or simply *product*) Z of two one-dimensional arrays (also called *vectors*) X, Y of the same length n, containing integer values. Although we start with two lists of integers, the result is a unique integer value computed with the following expression with the obvious meaning of the notation:

$$Z = X[1] \cdot Y[1] + X[2] \cdot Y[2] + \ldots + X[n] \cdot Y[n]. \tag{6.3}$$

Note that each row and each column of an adjacency matrix is in fact a vector of the same length n, so for example we can multiply row 3 by column

4 through the expression $\sum_{j=1}^{n}(A[3,j] \cdot A[j,4])$. With reference to the matrix A of Figure 6.2 we have:

$$A[3,1] \cdot A[1,4] + A[3,2] \cdot A[2,4] + \ldots + A[3,6] \cdot A[6,4]$$
$$= 0 \cdot 1 + 1 \cdot 0 + 0 \cdot 0 + 0 \cdot 0 + 1 \cdot 1 + 1 \cdot 1 = 2. \tag{6.4}$$

Perhaps surprisingly, the value 2 of the row-column product thus obtained has an important relation with graph G because it gives the number of paths of length two going from node 3 (the row index) to node 4 (the column index). But in fact this result tells us more. The computation shown in (6.4) indicates that the value 2 comes from the sum of the last two products $A[3,5] \cdot A[5,4]$ and $A[3,6] \cdot A[6,4]$ that have both value 1, and from this we know that the two paths going from node 3 to node 4 pass through nodes 5 and 6, respectively. Why is that so?

We need a little more algebra. The definition given above for vector multiplication is extended to two-dimensional $n \times n$ matrices X, Y whose product Z is a new $n \times n$ matrix, each element $Z[i,j]$ of which is an integer obtained as the product between the i-th row of X and the j-th column of Y. That is:

$$Z[i,j] = \sum_{k=1}^{n}(X[i,k] \cdot Y[k,j]). \tag{6.5}$$

Note now that the adjacency matrix A of a graph can be seen as a table of the paths of length one between any pair of nodes. In fact A contains only zeroes and ones since there may be at most one such a path for each pair of nodes i,j, namely the arc (i,j) itself. Consider the *square* of A, namely the matrix A^2 obtained multiplying A by itself; that is, $A^2[i,j] = \sum_{k=1}^{n}(A[i,k] \cdot A[k,j])$ for any pair i,j. If for a certain value of k we have $A[i,k] = A[k,j] = 1$, there is a path of length one from i to k and another path of length one from k to j, therefore there is a path of length two from i to j and a term $A[i,k] \cdot A[k,j] = 1$ appears in the expression of $A^2[i,j]$. If this happens for $h \geq 1$ different values of k, then h 1s sum up in the formula to give $A^2[i,j] = h$, that is the total number of paths of length two from i to j. As a consequence A^2 reports such numbers for each pair of nodes.

The squared matrix for the graph G is shown in Figure 6.2, where for example we have $A^2[3,4] = 2$ as already found in relation (6.4). We might observe that there is only one path of length two from node 1 back to itself, in fact the path 1-2-1; or that there is no path of length two leaving from node 4, as is obvious because there is no arc leaving from 4 hence there will be no path of any length. An immediate extension of this reasoning allows us to conclude that the cube of A, namely the matrix A^3 obtained multiplying A^2 by A, contains the number of all paths of length three for any pair of nodes, and in general A^k contains the number of all the paths of length k.[2] As we shall see

---

[2]Our discussion on the powers of the adjacency matrix is independent on the assumption that the arcs are directed. Therefore, for any $k \geq 1$, A^k contains the number of all the paths of length k both for directed or undirected graphs. For the latter graphs the values k even

this result applied to the graph of the Web is of paramount importance for determining the order in which the most popular search engine presents the answers to any given query.

6.2   The random growth of graphs

Having briefly surveyed the matrix representation and handling of graphs in a computer memory, we direct our attention to how a graph can be generated and how it may grow. Although this clearly depends on the field of application for which the graph has been taken as a mathematical model, some general rules always apply. Let us start with random growth, leading to the related question of what a random graph is. The analysis of such graphs plays only a minor role in computer networking, but it allows us to clarify some basic initial questions. The main actor in this field at the outset was Paul Erdős, a great Hungarian mathematician and a very peculiar character.[3]

Erdős worked in many fields of mathematics, collaborating with hundreds of colleagues around the world. In particular he developed the classical theory of random (undirected) graphs with his colleague Alfréd Rényi in 1959/60, initially based on the following simple process:

Random process 1

1. *start with an initial set of nodes;*

2. *insert arcs one by one, connecting pairs of nodes chosen at random.*

This process generates what is called a *random graph*. It is understood that if a pair of nodes is chosen several times, the corresponding arc is inserted only once. Note that the process refers to undirected graphs but can be extended to directed graphs in a straightforward way.

Before proceeding, it is important to be aware of exactly what it means to be dealing with random graphs. In the above process the number n of nodes is fixed a priori, while the number m of arcs to be inserted before the result is observed can be seen as an independent parameter as long as m does not exceed $n(n-1)/2$ (i.e., the maximum number of arcs in an undirected graph). If the graph is intended to model a real-life network it may be kept sparse, i.e., m may be bounded by a linear function of n. Note that the arcs are inserted

---

imply that the main diagonal of A^k contains all non zero elements because there is always a path going from a node back to itself traversing the same arc in two directions.

[3]The life and work of Paul Erdős merits further reading (see the bibliographic notes). He did not possess anything aside from a large suitcase containing some essential belongings. Not having a home he traveled continuously to visit friends with whom he developed all sorts of mathematical themes. He gave all his earnings to needy people or to young mathematicians as a reward for obtaining new results.

one by one in consecutive steps, so m is also the step number at which the process is monitored.

Aside from the two values n, m, the resulting graph G is unpredictable. So for example taking $n = 6$ and $m = 8$, the graph of Figure 6.1(a) might emerge from the random process with the same probability as any other graph with six nodes and eight arcs. On the other hand if we start by taking graph G on purpose this graph is clearly not random. The situation is the same as discussed for random sequences in the last chapter. The term random graph refers to any member of a statistical ensemble built with the repetition of the above process for given n, m: in fact we are interested in the properties that apply to the whole ensemble. In other words, the properties of a random graph are those of an ensemble of graphs.[4]

Given n and m, let the nodes and the steps of the process be numbered from 1 to n, and from 1 to m, respectively. The main parameter to be considered is the degree distribution $p(x, d)$ of each node x, that is, the probability that x has degree d. Clearly we assume that the node degrees are statistically distributed over the graph (or ensemble of graphs). Then we can pass from $p(x, d)$ to the *probability degree distribution* $P(d)$ of the whole graph defined as:

$$P(d) = \frac{1}{n} \sum_{x=1}^{n} p(x, d). \tag{6.6}$$

That is, $P(d)$ is the probability of encountering a node with degree d. If $m \geq n - 1$, as is necessary for the graph to be connected (although this may not be sufficient) the value of d of each node may span from zero to $n - 1$ and the mean value of d is given by:

$$\bar{d} = \sum_{d=0}^{n-1} d \cdot P(d). \tag{6.7}$$

This same value is linked to m by the relation: $\bar{d} = 2m/n$ already given above.

These considerations can be immediately extended to the in-degree and the out-degree of the nodes of a directed graph, thus defining $P(d_{in})$ and $P(d_{out})$ in a straightforward way. We also have $\bar{d}_{in} = \bar{d}_{out} = m/n$ as already found.

Let us now examine the function $P(d)$ in more depth. Although reflecting a property of the whole graph, this function is built as a composition of the degree probabilities of the single nodes, therefore it accounts for a "local" property without saying anything more on the graph structure. $P(d)$ gives strong statistical information on how many arcs are incident to each node, but not, for example, on which arcs belong to a cycle of a given length, or must be followed to reach another node in a certain number of "hops" (a term coming from computer networks). Nevertheless, knowing the probability degree distribution is of great importance for studying real-life networks, as well as being not too difficult to compute if some strict mathematical condi-

---

[4]For a more in-depth treatment of this point see the book of Dorogovstev and Mendes, 2003, mentioned in the bibliographic notes.

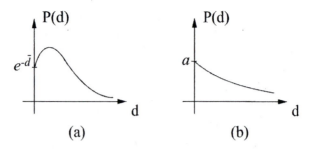

(a)   (b)

FIGURE 6.3: (a) Poisson distribution for random process 1. The value of d where $P(d)$ is maximum depends on $\bar{d}$ and increases with this value. $P(d) \to 0$ for $d \to \infty$. (b) Exponential degree distribution for random process 2. For $d \to \infty$ the curve goes to zero with exponential decay.

tions are satisfied. We then start from $P(d)$ in all the graph growth models. When coming to the Internet and the Web these mathematical studies supply a strong basis for understanding their development, although we will have to rely on experimental data because not all the side conditions are precisely known.

For the Erdős and Rényi's graph process with the arcs chosen perfectly at random, standard combinatorics allows to prove that $P(d)$ follows the *Poisson distribution*:

$$P(d) = e^{-\bar{d}} \cdot \bar{d}^d / d! \tag{6.8}$$

if the mean degree $\bar{d} = 2m/n$ is a constant, that is, if m is taken as a linear function of n.[5] The general behavior of function (6.8) is shown in Figure 6.3(a). Since d is a node degree, only the non-negative values of d carry meaning.

To get an idea why the function has this shape, recall from Chapter 2 that the factorial of d can be expressed in the form $d! \sim \sqrt{2\pi d}(d/e)^d$ (Stirling's approximation). Removing all constant terms from relation (6.8) we have:

$$P(d) \propto \bar{d}^d/(\sqrt{d}(d/e)^d) = (\bar{d} \cdot e)^d/d^{d+\frac{1}{2}} = a^d/d^{d+\frac{1}{2}} \tag{6.9}$$

where $a = (\bar{d} \cdot e)$ is a constant, and the symbol $\propto$ denotes proportionality. For small values of d (essentially for $d < a$) the numerator a^d grows faster than the denominator $d^{d+\frac{1}{2}}$ and the function grows. As d increases, the two terms first balance and the function reaches a maximum, and then the denominator definitely wins and $P(d)$ goes to zero. In fact we have $a^d/d^{d+\frac{1}{2}} < a^d/d^d = (a/d)^d$, with an exponential decay for $d \to \infty$ as the base a/d becomes less than one.

---

[5]For a mathematical proof of the expression (6.7), and of most of the expressions of this chapter, see for example Dorogovstev and Mendes, *op. cit.*

The analysis of the Erdős and Rényi's process leading to the distribution (6.8) is strictly valid for increasing n (and fixed $\bar{d}$), besides being related to a continuous function while d can actually take up only integer values. Nevertheless it is indicative of what one may expect from this growing mechanism. Now we've understood this basic model, at least in broad terms, we can examine a new model that is closer to what happens in real growing networks like the Internet and the Web.

Unlike the previous model, both nodes and arcs can now grow in number. The mathematical approach is the same as before, but the resulting form of $P(d)$ is completely different. Referring again to undirected graphs we pose:

Random process 2

1. *start with an initial node and fix an integer constant $k \geq 1$;*

2. *proceed with consecutive steps: at each step insert a new node and k arcs connecting pairs of nodes chosen at random.*[6]

It can be easily seen that, from step $i = k + 1$ on, the total number of arcs is $m = nk - k(k - 1)/2$. Since k is a constant, as the process goes on we have $m \sim nk$, hence $\bar{d} \sim k$.

This process is too simple for modeling exactly most real life phenomena where, for example, arcs are not compelled to grow in batches of the same size k for any new node; not to mention that both nodes and arcs may increase and occasionally also decrease in number. Again we take a simplified version for which a simple mathematical analysis is possible, keeping in mind that the general behavior of similar processes would be quite similar. And again we study $P(d)$ as a continuous function although d must actually take integer values.

Standard mathematical analysis shows that the function $P(d)$ for random process 2 follows the *exponential distribution*:

$$P(d) = a \cdot e^{-d/\bar{d}} \tag{6.10}$$

where a is a constant depending on k.[7] The well known behavior of this function is shown in Figure 6.3(b). Let us see why the two functions (6.8) and (6.10) are so different.

First note that, for $d \to \infty$, the Poisson distribution goes to zero much faster than the exponential distribution because both functions exhibit an exponential decay, respectively with $(a/d)^d$ and $(1/e)^d$ where both bases are < 1, but in the first one the base also decreases with d while in the second the base is a constant. Recalling that $P(d)$ is the probability of encountering a node

---

[6]In the first $i \leq k$ steps only $i - 1$ arcs can be inserted, connecting the new node with the existing $i - 1$ nodes. From step $i = k + 1$ on, k arcs are added at any step.

[7]For a mathematical proof of the expression (6.10) see again Dorogovstev and Mendes, *op. cit.* sections 2.1 and 5.1.

with degree d, it follows that for each large value of d the exponential distribution has many more nodes with that degree. In fact, while in the Poisson distribution the degrees of the different nodes tend to gather "democratically" around a mean value roughly coincident with the maximum of the function, that is most nodes have a degree close to the average and very few of them have a high degree, in the exponential distribution many more nodes have a higher value of d. Why should that be the case?

Giving an intuitive answer is not that difficult. In process 1, when the arcs enter the game, the nodes are already there, so each arc may be attached to each node with the same probability. As a consequence all the nodes have the same chance of reaching any given degree. In process 2, however, if node x is added to the graph before node y, x has a higher probability of being targeted by some arcs for the sole reason that it is in the game longer. Therefore the degree distribution generated by process 2 is unbalanced, with "older" nodes having on average a higher degree.

To learn more about process 2, which is closely related to many real life networks, it is convenient to number the nodes according to the step in which they appear (i.e., node x is inserted in the graph in step x). Then, in the statistical ensemble of graphs defined by the process, $p(x, d, i)$ is the probability that node x has degree d at step i, with $i \geq x$; and the total degree distribution at step i is given by: $P(d, i) = \frac{1}{i} \sum_{x=1}^{i} p(x, d, i)$. Clearly we have $P(d, n) = P(d)$, and the exponential relation (6.10) follows.

We also define $d(x, i)$ as the mean degree of node x at step i, with $i \geq x$. Its mean value $\bar{d}(x, i)$ is an interesting element in understanding how the graph grows because it can be proved that:

$$\bar{d}(x, i) \propto \log(i/x). \tag{6.11}$$

That is, the node degree increases with the ratio of i over x (older nodes have a higher degree, as expected), but such an increase is moderate since it follows a logarithmic function.

What is interesting, however, is that processes 1 and 2 give rise to networks with a small distance between arbitrary nodes. In fact in both processes the mean value of the shortest path can be evaluated as:

$$\bar{l} \simeq \log n / \log \bar{d}. \tag{6.12}$$

That is $\bar{l}$ scales logarithmically with n so that the mean distance between nodes remains short even in a huge network.[8]

Finally, remember that processes 1 and 2 have been examined for undirected graphs but their extension to directed graphs is straightforward, leading to similar results for the functions $P(d_{in})$ and $P(d_{out})$.

---

[8]Logarithms with different bases differ only by a multiplicative constant as explained in Chapter 3, relation (3.1). Therefore the value of $\bar{l}$ in relation (6.12) is independent of the base, as far as the same base is chosen for the two logarithms.

6.3 Power laws: the rich get richer

Although randomness plays an important role in the growth of real life networks including the Internet and the Web, a completely different growth model may be even more relevant. As we will see now, this phenomenon was originally observed in the field of economics, and only later made its way into other sciences.

The story starts at the end of the nineteenth century with Vilfredo Pareto and his theories on political economy.[9] In 1896 Pareto presented the "curve of income" based on an accurate statistical survey on the income distribution in different European countries, particularly England and Prussia where available data were more reliable. This curve, drawn in Figure 6.4(a), shows the number y of persons with income at least x. The best mathematical fit for it was given by Pareto in the form:

$$y = b(x + a)^{-\gamma} \tag{6.13}$$

where a, b, and γ are positive constants. With the terminology used today equation (6.13) is a *power law* since y is expressed as a power of x with the addition of some constant terms. Studying networks we will find power laws quite often. In all cases the exponent is negative hence the curve is convex and y decreases for increasing x.

One could have expected that the constants a, b, γ depended on the country or on the year in which the phenomenon was observed, but Pareto's statistics showed that their values were quite similar in all countries for which data were available, and remained almost unchanged in a time window of over forty years. In particular the value of a was always extremely small, and the value of γ spanned from 1.89 to 1.60. As Pareto always stressed, the basic "shape" of the curve (in fact, the power law form) characterizes the distribution of wealth everywhere, even if some changes in the constants can induce minor deformations. Before explaining what all this has to do with the Internet, let us examine the meaning and the consequences of such a mathematical behavior.

At a first glance the curve shows that a very large number y of persons have a very low income at most x and only a few persons have a very large income. This fact can be characterized with an elementary mathematical analysis, us-

---

[9]The son of an Italian political refugee, Vilfredo Pareto grew up in Paris. His family was eventually allowed to return to Italy were he graduated in engineering and then reached a very high position as an executive in industry, although he was openly a socialist and culturally an anarchist. Later he left industry to become a professor of Political Economy in the Swiss University of Lausanne where he wrote his famous *Cours d'économie politique* (Course of Political Economy) in which economics was approached on quantitative grounds, a novelty for those times. His political beliefs were so inflexible that he even refused a life seat in the Italian Senate.

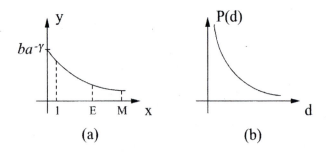

FIGURE 6.4: (a) Pareto's income distribution. The curve starts at a high value on the y axis since a is very small. (b) Power law of node degree distribution for a typical network growth.

ing once again a continuous function (6.13) to describe a discrete phenomenon (number of persons versus amount of income) in its general terms. Confining the curve into a finite window, say $1 \leq x \leq M$, to represent an actual span of income; and letting E be the mean value of the income over that span; it can be proved that the persons earning less than E are many more than one half of the total population independently of the values of a and b. That, is the poor are a vast majority.

In terms of political economics this fact can be simply explained. People with a very low income can barely survive and tend to stay in their condition forever. If their personal income increases there is some room for savings, and further income comes from the "interest" on the capital, whatever interest means. As the income becomes even larger people start acquiring further sources of revenue and become more and more wealthy. Following a popular saying: "the rich get richer." Economic data are presented today in much more sophisticated forms compared to Pareto's curve, but a basic fact still holds. From the poorest to the richest countries, a small percent of the population owns a large percentage of the total wealth.

In the decades that followed Pareto's studies, other power laws were applied to the mathematical description of different phenomena, many of which related to human affairs. Starting from the 1930s a significant role in empirical statistics was played by studies on the frequency of the words in natural language utterances, mainly due to the Harvard linguist George Kingsley Zipf. The original *Zipf's law* stated that, if the words of a sufficiently rich linguistic corpus are ordered according to their decreasing frequencies, the following relation (approximately) holds:

$$f(w_i) \propto i^{-1}$$

where w_i is the word in the i-th position (*rank*) of the ordered list, and $f(w_i)$ is its frequency. That is, the frequency of any word w_i is inversely proportional to its rank. Taking English as an example, the Zipf's law states that the most frequent word "the" occurs approximately twice as often as the second word "of," three times as often as the third word "and," etc., as is in fact observed in relevant collections of English sentences. The law was then extended by raising the rank i to a distribution exponent different from -1, and examining different fields in which it may hold.

Note that power laws are not the only mathematical expressions for which a small portion of a "population" accounts for a high share of the total "wealth," whatever the terms in quotes represent. A similar effect occurs if the wealth is distributed according to an exponential law where the independent variable x is at exponent (with minus sign), instead of a power law where x is in the base; however, there is a crucial difference that will be explained below.

It is now time to discuss how power laws appear in graphs. In fact, although there has been an excessive pursuit to discover power law distributions in all sorts of data collections, sometimes exaggerating their validity to unduly support different claims, such distributions unquestionably play a central role in the growth of networks. Consider the following growth process where, at any step, the nodes with the highest degree have a better chance to increase their degree (i.e., the rich get richer principle applies). Referring again to an undirected graph we pose:

Preferential attaching process 3

1. *start with an initial node 1;*

2. *proceed with consecutive steps: at each step i insert a new node i and a new arc connecting i to an existing node x chosen with probability p proportional to the current degree of x (i.e., $p \propto d(x, i)$).*

This process, proposed by Barabási and Albert in 1999, is one of the bases of the whole theory of network growth, and gives rise to a so called "citation graph." Standard mathematical analysis shows that the function $P(d)$ emerging from process 3 has the continuous power law form:

$$P(d) \propto d^{-3}. \tag{6.14}$$

Compared to Pareto's curve of Figure 6.4(a) the function (6.14) goes to ∞ for d going to zero. However the new curve has a meaning only for $d \geq 1$ because by construction all the vertices have at least one incident arc. Furthermore the exponent -3 in relation (6.14) is larger in absolute value than the exponent $-\gamma$ of (6.13) so the new curve is closer to the x axis.

A little caution is in order here. Equation (6.13) gives the number of people with income $\geq x$ (not exactly x), while equation (6.14) gives the probability of finding a vertex with degree equal to d. We can easily transform the latter equation to give a *cumulative* distribution $P_c(d)$, i.e., the probability of finding a vertex of degree $\geq d$. In fact $P_c(d)$ is simply the integral of $P(d)$ from d to ∞, hence we have $P_c(d) \propto d^{-2}$. In the next chapter, dealing with the Internet and the Web, we will see that cumulative distributions are often more significant than the others.

Many variations of process 3 have been proposed in order to make it more suitable for modelling real networks. Major extensions allow several arcs to enter at each step, although their number must be kept constant along the process to permit a reasonable mathematical analysis. Moreover these arcs may be connected to the nodes according to a mixture of preferential and random attaching, as in fact happens in the Internet and the Web. Finally there is no need that a new node x is immediately connected to the others, although this requires a little caution (see below).

All these models must be studied in the specialized literature. We indicate only a very simple extension of process 3 that gives a significant account of what can be expected from the others. Namely:

Preferential and random process 4

1. *start with an initial node 1;*

2. *proceed with consecutive steps: at each step i insert a new node i and a new arc connecting two existing nodes x, y chosen with probabilities $p_x \propto d(x, i) + a$ and $p_y \propto d(y, i) + a$, with $a > 0$ constant.*

Process 4 introduces a mixture of preferential and random attachment, the latter through the additive constant a. The greater a is, the less preferential is the process. Although possibly very small, a cannot be zero because each new node i enters the graph with degree zero and, for $a = 0$, could never be attached to the others. Asymptotic analysis shows that for large values of d the function $P(d)$ emerging from process 4 has the form:

$$P(d) \propto d^{-\gamma} \tag{6.15}$$

with $\gamma = (2 + a/2)$. To stay close to what happens in many real networks a must be chosen in the interval $(0 - 2]$ so γ has a value in $(2 - 3]$.

All the common variations of process 4 end up with a power law for $P(d)$ whose exponent and other constants are a function of the various parameters of the process such as the number of arcs introduced at each step, or their distinction between preferential and random arcs, or even the number of arcs that are taken off the graph at certain steps. All these paramenters are reflected in the graphical representation of the function, which in all cases has the shape of Figure 6.4(b). In a later chapter we shall see that several other features of the Internet and the Web, and of other real networks, are governed by power laws with most exponents between -3 and -2.

We can also consider the value of $\bar{d}(x, i)$ (mean degree of node x at step i) as we did for the exponential distribution. For process 4 and its variations we get in general something like:

$$\bar{d}(x, i) \sim (i/x)^{\beta}, \tag{6.16}$$

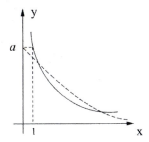

FIGURE 6.5: Asymptotic comparison between random attachment (exponential law in dashed line) and preferential attachment (power law in solid line).

with $\beta < 1$. That is, the mean node degree increases with the ratio of i over x as expected, and increases much faster than in the exponential distribution (relation (6.11)) where the variation is limited to a logarithmic growth.

Even for process 3 and for its extensions, the mean distance between nodes is small, generally scaling with n as:

$$\bar{l} \sim \log n / \log \log n. \tag{6.17}$$

We have now the basic information required to compare an exponential distribution with a power law distribution. Figure 6.5 shows the two functions:

$$y = a\, e^{-bx}, \quad y = a\, x^{-\gamma} \tag{6.18}$$

where a, b, γ are constants.

The points where the two curves intersect the x, y axes, or intersect each other, depend on the values of a, b, γ, but the general shape of the curves is independent of these parameters. The exponential law may lay above the power law in an intermediate interval of the x coordinate, but stays always below the power law for small and large values of x. Furthermore for $x \to \infty$ the exponential law goes to zero (by definition) with exponential decay, while the power law goes to zero much more slowly. The value of the mean degree of a node as a function of the instant when the node enters the game, indicated in relations (6.11) and (6.16) for the two distributions, confirm the abundance of nodes with high degree in the latter case. Power laws are said to have a "fat tail."

An interesting study of the two functions (6.18) can be performed if we take the natural logarithm of both sides of the equality, thus obtaining:[10]

$$ln\, y = ln\, a - b\, x, \quad ln\, y = ln\, a - \gamma\, ln\, x. \tag{6.19}$$

---

[10] Recall that $ln\,(a \cdot b) = ln\, a + ln\, b$; $ln\, a^b = b\, ln\, a$; and $ln\, e = 1$.

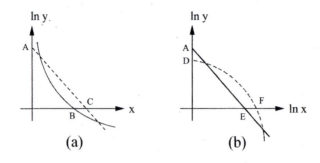

FIGURE 6.6: Exponential behavior (dashed) versus power law (solid), in semi-logarithmic scale (a), and in logarithmic scale (b). The intersections are: $A = \ln a$, $B = a\,e^{-\gamma}$, $C = \ln a/b$, $D = \ln a - 1$, $E = \ln a/\gamma$, $F = \ln\ln a - \ln, b$.

Now plot the two functions on the new axes $x, \ln y$ (semi-logarithmic scale: Figure 6.6(a)), and $\ln x, \ln y$ (logarithmic scale: Figure 6.6(b)). In the first plane the exponential function corresponds to a straight line with slope $-b$. In the second plane the power law corresponds to a straight line with slope $-\gamma$. As noted in footnote 8 the shape of these curves is essentially independent of the base chosen for the logarithms. In practical applications base 10 is mostly used.

The shape of the curves in Figure 6.6 is very important for evaluating experimental results. In fact data coming from a random attachment experiment, or from a preferential attachment experiment, tend to cluster along a straight line respectively in a plane $x, \log y$, or in a plane $\log x, \log y$. So a mere examination of the data distribution on the plane may reveal, at least approximately, the nature of the experiment.

Those who have reached the end of this rather heavy chapter may look at what follows with some relief. Reading the remaining chapters is going to be a smoother and more pleasant ride.

Bibliographic notes

Graph theory is the subject of countless books. To mention just one classic, still widely available: Berge, C. 1962. *The theory of graphs and its applications.* John Wiley & Sons, New York. The field of networks is more recent, and it is advisable to start with something not too difficult. As indicated in Chapter 1, an introduction directed to a general public is contained in: Barabási, A.L. 2002. *Linked: The New Science of Networks.* Perseus Publishing, Cambridge, MA, whose reading requires a modest knowledge of mathematics. To go deeper into the subject consider for example: Bornholdt, S. and H.G. Schuster, Editors. 2002. *Handbook of Graphs and Networks.* Wiley–VCH, Berlin.

The mathematical properties of graphs related to network modeling are scholarly analyzed in the book: Dorogovstev, S.N. and J.F.F. Mendes. 2003. *Evolution of Networks*. Oxford University Press, where the interested reader can find a complete mathematical justification of many of the properties of networks described in the present book. A simpler treatment of similar subjects can be found in: Barrat, A., M. Barthélemy, and A. Vespignani. 2008. *Dynamic Processes on Complex Networks*. Cambridge University Press.

The fascinating story of Paul Erdős can be read in the very pleasant book: Hoffman, P. 1988. *The Man Who Loved Numbers: The Story of Paul Erdős and the Search for Mathematical Truth*. London: Fourth Estate Ltd.

The *Oeuvres complètes de Vilfredo Pareto* (The Complete Works of Vilfredo Pareto) have been published in Switzerland in thirty volumes, all written in French. The major biographies are also in French. An accurate summary in English of Pareto's life and work can be found in Wikipedia.

Chapter 7

Giant components, small worlds, fat tails, and the Internet

How networks appear in the real world, and how they can be studied with a reasonable mixture of mathematics and observations.

One of the greatest consolations of this world is friendship, and one of the pleasures of friendship is to have someone to whom we may entrust a secret. Now, friends are not divided into pairs, as husband and wife: everybody generally speaking, has more than one; and this forms a chain of which no one can find the first link. When, then, a friend meets with an opportunity of depositing a secret in the breast of another, he, in his turn, seeks to share in the same pleasure. He is entreated, to be sure, to say nothing to anybody; and such a condition, if taken in the strict sense of the words, would immediately cut short the chain of these gratifications: but general practice has determined that it only forbids the entrusting of the secret to everybody except one equally confidential friend, imposing upon him, of course, the same conditions. Thus, from confidential friend to confidential friend, the secret threads its way along this immense chain, until, at last, it reaches the ear of him or them whom the first speaker exactly intended it should never reach. However, it would, generally, take a long time on the way, if everybody had but two friends, the one who tells him, and the one to whom he repeats it with the injunction of silence. But there are some highly favoured men who reckon these blessings by the hundred, and when the secret comes into the hands of one of these, the circles multiply so rapidly that it is no longer possible to pursue them.[1]

So if at the next step one of these "highly favoured men" tells the secret to one hundred friends, who are probably also highly favored, the secret is deposited in the breasts of $100^2 = 10,000$ new custodians, and the process

---

[1]Manzoni 1827, Ch. XI. The stylish Italian novel *The betrothed* is considered a masterpiece of world literature. It may well be that these authors have inserted the citation because Manzoni's novel is an inescapable part of their personal culture. In any case, this description of secret spreading predicted the chain email almost two centuries in advance.

117

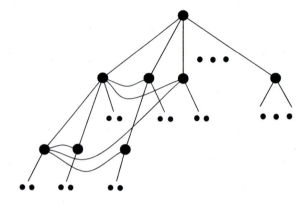

FIGURE 7.1: A graph of secret sharing with connections between mutual friends.

goes on exponentially along a communication tree where the number of nodes is multiplied by 100 at each level.

Chances are, however, that the 10,000 custodians are not all distinct, according to the universal law that "a friend of my friend is also my friend." The set of custodians then grows in the form of a complex communication network as shown in Figure 7.1. Unlike the tree of exponential growth, the network is highly clustered, with groups of friends all linked to one another, and short loops appear. On the other hand the exponential growth cannot go on for too long, because in a few steps, the total number of custodians would exceed the population of the globe. Either the tree stops evolving, or closes onto itself forming long loops.

In Chapter 6 we have seen how networks can grow according to different mathematical rules, with random or preferential linking, or with a mixture of the two. Clearly different networks require different algorithms and the same problem may be easy on one and difficult on another. Which parameters are then relevant for network study? Let us approach this by trying to answer a natural question: what happens in the real world?

7.1   The emergence of giant components

In their original work on random graphs, growing with successive attachments of arcs to a given set of nodes, Erdős and Rényi discovered an interesting phenomenon. Their graphs showed a sudden change (today called a "phase transition") in node connectedness for increasing values of the mean node degree $\bar{d}$. In the initial steps of the growing process, that is when there are just a few arcs in the graph and the value of $\bar{d}$ is small, most nodes are disconnected from one another and the graph actually consists of a great number of independent connected components. Increasing the number of arcs, these components start merging to form bigger components; until, when the value $\bar{d} = 1$ is reached, a very large connected component emerges whose nodes form a sizable subset of the total.

More precisely, if we repeat the process for an increasing number of nodes n, all initial components have a size s with $\frac{s}{n} \to 0$ for $n \to \infty$. Around $\bar{d} = 1$, however, a *giant* (connected) *component* appears whose size s is a finite fraction of n, i.e., $\frac{s}{n}$ is a positive constant for increasing n. The graph is then composed of one giant component plus a collection of components of negligible size. By adding new edges, the latter components tend to merge with the giant one, until the whole graph is connected. With a little injustice we will essentially direct our attention to the giant component only. For example the mean length $\bar{l}$ of the shortest path between all pairs of nodes implicitly refers to the nodes of the giant component. Note that $\bar{d} = 1$ implies that each node has on average just one incident arc, so studying graphs for $\bar{d} \geq 1$ is reasonable on practical grounds.

Erdős and Rényi's model was an abstract one, and the birth of a giant component was demonstrated in mathematical terms. In fact this phenomenon occurs in most real networks, although we may not be able to predict the values of the parameters involved in the phase transition. For example, if the nodes of a network are the freshmen of a college, and an arc is formed whenever a new acquaintance arises, the corresponding graph is initially formed of many small components, but at the end of the academic year the graph is dense and a giant component is certainly present, leaving out only small groups of shy people, or of fellows with narrow common interests. Unlike in the Erdős and Rényi's model this process is essentially non random since acquaintances may be due to personal attraction, or to interest or taste sharing. Still one (and only one) giant component will arise.

So far so good. But the phenomenon becomes much more complex if the graph is directed. A much larger number of arcs is needed to form directed paths between all the nodes of a component and the global structure of the connections becomes more interesting. For a moment, take all the directed arcs as undirected. The giant component so observed is said to be *weakly connected*, and is called GW for *giant weak*, to indicate that many connections would disappear on restoring the arc orientation. As before, the graph is divided into GW plus several independent small components, and again we direct our attention to GW.

Restoring the arc orientation, GW takes the form indicated in Figure 7.2. If the number of edges is large enough, the three subsets GS, GI, and GO have a size proportional to n, hence they are properly called giant. The core GS, for *giant strong*, is the largest connected subset of GW. Each node in GS can be reached from any other node, that is GS is a connected component

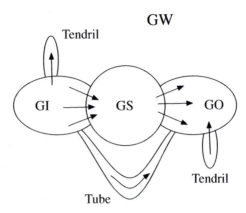

FIGURE 7.2: The structure of the weak giant component GW in a directed graph. The core GS is the largest connected component of the whole graph.

of the directed graph. GI, for *giant in*, contains all the nodes from which a node in GS can be reached, but non vice-versa (otherwise the node would be in GS). Conversely GO, for *giant out*, contains all the nodes that can be reached by the nodes in GS, but non vice-versa. The remaining nodes of GW are clustered into *tendrils* and *tubes*. Tendril nodes can be reached from GI, but cannot reach nodes outside the tendril; or can reach GO, but cannot be reached from outside the tendril. Tube nodes form a bridge connecting GI and GO without passing through GS.

As we shall see, the structure of Figure 7.2 is found in the World Wide Web represented as a directed graph. Taking our college freshmen again, we may construct a directed graph of telephone calls where arcs are directed from the caller to the receiver. The fellows in GS make and receive many calls. Those in GI try to be friendly without great success. Those in GO are very much sought for but tend to keep to themselves. The reason why this structure is found in the Web is of course very different.

Like random graphs, most non random networks tend to suddenly form one giant component when the number of arcs reaches a certain limit, unless special conditions occur.[2] This is remarkable since most of the other characteristics of the network, for example the degree distribution seen in Chapter 6, are strongly influenced by the formation process.

In a random graph all possible arcs have the same probability of occurring independently of the arcs already present, while in most real-life networks, if two nodes x, y are connected to another node z, the probability that x and

---

[2]For example the freshmen of two colleges located in different states, if taken as a whole, tend to generate two giant components of half size, one for each college. However both components have a size proportional to the total number of nodes, hence are giant.

y are also connected by an arc is generally much higher than casting arcs at random ("a friend of my friend is also my friend"). There is a vast literature on this clustering phenomenon, or in general on the local density of a network given its global density. In an undirected graph a node x of degree d has d neighbors that may share up to $s = d(d-1)/2$ arcs. If c arcs actually link the neighbors of x, the standard *clustering coefficient* of x is defined as:

$$C = c/s = 2c/d(d-1), \tag{7.1}$$

and the mean value $\bar{C}$ of C can be computed as a parameter of the whole network. For directed graphs, relation (7.1) can be immediately extended.

For a wealth of social networks, biological networks, neural networks, distribution networks, and for communication networks like the Internet and the Web, the value of $\bar{C}$ has been shown to be much higher than that of a random network with the same number of nodes and arcs. In all these cases, two neighbors of a given node have a rather high probability of being neighbors of one another, thereby keeping the value of $\bar{C}$ high, since this parameter indicates the density of loops of length three. More generally two nodes at a short distance from a given node may be neighbors of one another thus contributing to local clustering, as happens for example in the graph of Figure 7.1. Global clustering can be measured with a proper extension of relation (7.1) to indicate the density of loops of any length.

In a random graph, however, each node has probability $d/(n-1)$ that one of its d incident arcs is in fact connected to one of the other $n-1$ nodes. Recalling that for increasing n the value of d must be kept constant in practice (otherwise the "size" of the nodes would be unbounded), the above probability is asymptotically vanishing and the value of c in relation (7.1) goes to zero. Then $\bar{C} \to 0$ for $n \to \infty$. Random networks are practically free of clustering, that is they tend to have a tree-like structure in the surrounding of each node and loops are long. The giant connected component present in all networks has an internal structure that depends on the growing process. Other interesting properties will be found in the next section.

7.2   The perception of small worlds

Manzoni's secret sharing suggests that all of the approximately ten billion people that form human society would be reached by a secret in a few steps, except possibly for small groups of asocial individuals that remain outside the giant component. Apparently this is not that far from the truth, as the network of personal acquaintances between humans seems to have a giant connected component where the mean distance between any pair of nodes is

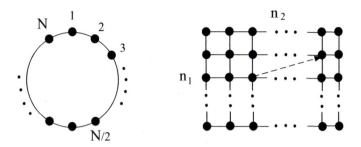

FIGURE 7.3: Two non small-world graphs: a ring and a two-dimensional grid. A dashed "long-range contact" has been added to the latter as in Kleinberg's model.

quite small. This is what is known as the *small-world effect* in the field of social networks.

These studies started in the 1960s with a set of experiments conducted by the psychologist Stanley Milgram at Harvard University. The most relevant experiment for the field of networks is so well known that we will recall it only very briefly. Milgram asked a random group of people in Nebraska to send a letter to a specific person in Boston passing through a chain of people that at each step the sender personally knew. Many letters got lost, but those that reached the Bostonian addressee had made an average of six hops. The expression "six degrees of separation" then passed into popular language, to indicate that six (or, in any case, a very small number of) steps of acquaintance separate any two of us: from Mahmud Ahmadinejad to Angelina Jolie; from Angela Merkel to Tiger Woods.

From a mathematical point of view the small-world effect arises when the mean distance $\bar{l}$ between nodes is at most polynomial in $\log(N)$, where N is the number of nodes. Not "six" as found by Milgram; not even a higher constant, that would be demanding too much as N goes to infinity; but a big reduction anyway. Note that we are referring to $\bar{l}$, that is, to the mean length of the shortest path between any two nodes, and not to the *diameter* δ of the graph defined as the length of the *longest* shortest path (i.e., the path connecting two vertices at maximal distance).[3]

In Chapter 6 we have seen that networks growing randomly or with preferential linking are small worlds. Of course this result comes from a mathematical abstraction, but all social or communication networks examined in the last decades experimentally exhibit the same property. So, to start with,

---

[3]There is some confusion in the literature on using $\bar{l}$ or δ to define small worlds. This may be due to the fact that in a random graph both parameters are logarithmic in N. We refer to $\bar{l}$. As $\bar{l}$ is a mean value, it may well be that the actual distance between Angela Merkel and Tiger Woods is much more than six.

which graphs *are not* small worlds? Typical examples in computing are *rings* of processors used in small local area networks, and *regular grids* of CPUs used in parallel processing: see Figure 7.3. In a ring of N nodes we can easily find:

$$\delta = \lfloor N/2 \rfloor, \quad \bar{l} \sim N/4. \tag{7.2}$$

In a two-dimensional grid of $N = n_1 \times n_2$ nodes we have:

$$\delta = n_1 + n_2 - 2, \quad \bar{l} \sim (n_1 + n_2)/2, \tag{7.3}$$

both proportional to the square root of N if n_1 and n_2 are of the same order of magnitude. In general in a d-dimensional grid both δ and $\bar{l}$ are of order $O(N^{1/d})$. On the other side of this spectrum stand complete graphs with $\delta = \bar{l} = 1$.

Comparing random graphs and grids we see that two of their characteristics are opposite. In random graphs the value of $\bar{l}$ is very low and the loops are long (low clustering), while in grids the value of $\bar{l}$ is very high and the loops are short (high clustering). In 1998 Watts and Strogatz suggested how to reconcile the two models into one that exhibits low $\bar{l}$ and high clustering. For example starting from a highly regular and highly clustered network like a grid, add a small number of arcs chosen uniformly at random. The clustering remains, however, a few new random arcs are sufficient to reduce the value of $\bar{l}$ dramatically, and tuning the parameters this value becomes logarithmic. This process creates a small world with high clustering. A standard example is a set of tightly connected communities (e.g., freshmen of different colleges), with the communities initially disconnected from one another. A few random connections between nodes of different communities are sufficient to link any pair of nodes with just a few hops, part on the new arcs and part inside the communities.

Speculating on Milgram's experiment and on Watts and Storgatz models, Jon Kleinberg discovered some interesting properties of *decentralized algorithms* for the detection of short paths in a network. These algorithms only make use of some local information stored at each node, as in fact was attained by the participants in Milgram's experiment. The problem is put in strongly mathematical terms, focusing on a generalization of an $n \times n$ grid model on which three parameters p, q, r are imposed. Each node u now has a direct connection with every other node v within a distance p (i.e., p is the number of grid steps between u and v). The nodes in this neighborhood are the *local contacts* of u. In particular for $p = 1$ the local contacts are the four nodes surrounding u in the grid. In addition u has q *long-range contacts* determined by q independent random trials. The probability that u chooses v as one of these contacts is proportional to $d(u, v)^{-r}$, where $d(u, v)$ is the number of grid steps between u and v. In particular for $q = 1$ there is only one long-range contact per node (see Figure 7.3, letting $n_1 = n_2 = n$).

Kleinberg proved two interesting facts about this model. For $p = q = 1$, i.e., if each node u has only four local contacts and one long-range contact; and for $r = 2$, i.e., the probability of connecting u with the long-range contact v is proportional to $d(u, v)^{-2}$; one can apply a very simple decentralized routing algorithm. To send a message to a target node t, the original source node u, and then any successive message holder, sends the message to its contact that is closest to t. Kleinberg proved that this algorithm requires an expected number of hops of order $O((\log n)^2)$, that is, a delivery time exponentially smaller than the number $N = n^2$ of grid nodes. Not only does the grid become a small world with the addition of one long-range contact per node, but also each node is able to send the message along a short path to a given target solely on the basis of limited local knowledge.

The second result is also impressive. For $r \neq 2$, any decentralized routing algorithm in the family of grids requires a delivery time polynomial in n instead of $\log(n)$: i.e., it is exponentially slower than the algorithm above. For example for $r = 0$ the long-range connections are chosen uniformly at random, since the probability of being chosen has the same value $1/N_l$ for all the possible N_l long-range contacts. Essentially this is the model of Watts and Storgatz. Short connecting chains also exist in this case, but no decentralized algorithm can detect them efficiently.

In conclusion a combination of randomness and local clustering are two basic ingredients to build a small world close to what is observed in practice.

7.3 Fat tails

In this world, smaller animals tend to be in the vast majority. There are many more mosquitoes than humans, many more humans than bulls, and only a few hippos. And the property scales: small insects are the majority of all insects, small mammals are the majority of all mammals, etc.

Many artificial or natural phenomena exhibit a similar property, at least in the "tail" of the distribution. On September 11, 2001, a dreadful terrorist attack hit the United States of America taking the life of about 3,000 people. On January 12, 2010, a tremendous earthquake hit Haiti and the victims were about 30,000. Unfortunately many other terrorist attacks stain the world with blood and many other earthquakes make our lives insecure. But there are only a few large ones, medium size attacks or quakes are more numerous, and the minor ones form the vast majority.

In America there are many small towns, a fair number are of medium size, but there are only a few huge metropolis.[4] Most U.S. airports have only a few flights per day, some are better served, but there are only very few really big

---

[4]This phenomenon of urban concentration was already observed by Felix Auerbach in 1913, between Pareto's and Zipf's studies.

hubs. Most books appear very rarely in bibliographies, some are referenced more often, while a small minority are mentioned almost everywhere. Examples are countless. Log files for mail servers show that most addresses have a limited number of shared messages, some have more, and a few have a huge number of connections. In the distribution of links in the World Wide Web a vast majority of pages have an extremely limited number of incoming links, some are more pointed to, and only a few are very popular.

Most of these phenomena are ruled by a power law, as found by Pareto for the distribution of income and by Zipf for the frequency of words (see section 6.3 in the previous chapter). From the number y of earthquakes of a given magnitude x, to the number y of Web pages with x incoming links, the mathematical expression of the relation is approximately:

$$y = a\,x^{-\gamma} \tag{7.4}$$

with a and γ positive constants: a power law. There are examples of diverse origin: from biology to social sciences, from communication to transport. And from the Internet and the Web as we shall discuss later.

As we have seen in Chapter 6, relation (7.4) can be represented in a $\log x$, $\log y$ plane as a straight line with negative slope γ (Figure 6.6(b)). And in fact the discovery of power laws in artificial or natural phenomena has generally come out of observations of experimental data plotted in logarithmic scale. Other properties, however, are peculiar to power laws and are important for their consequences.

Compared to the Poisson distribution typical of random phenomena, a power law has no peak around the average value of y (see Figures 6.3(a) and 6.4(b) of Chapter 6). Furthermore power laws go to zero quite slowly for increasing x, so that the elements with a very large value of x are a minority but their probability of showing up is non-negligible. They disappear only if x reaches a point of "physiological" cut-off (an earthquake of magnitude 10 in the Richter scale; an airport with as many connections as the total number of existing airports). This actually means that "the rich" at the far right of the x axis are sufficiently numerous to get most of the total "wealth," as was demonstrated by Pareto in economic terms over one century ago, and is unfortunately still true today as far as the distribution of income is concerned. This property is referred to as a *fat* (or *heavy*) *tail* of the distribution. In these phenomena a mean value of x is not particularly significant, since the distribution is completely unbalanced to the right.

Another fundamental property of power laws, tightly connected with those previously discussed, is that they *scale*, and in fact they are the only mathematical laws with this property. In mathematical terms this means that multiplying x by a constant c (or *scaling* x by c) causes a proportional change (scaling) of y. That is, a function $y = f(x)$ scales if $f(cx) \propto f(x)$. And indeed for a power law $f(x) = a\,x^{-\gamma}$ we have:

$$f(cx) = a(cx)^{-\gamma} = c^{-\gamma}ax^{-\gamma} = c^{-\gamma}f(x). \tag{7.5}$$

Note that the exponent of x is negative for most of the power laws that we consider here, that is, the function is *convex* in a mathematical sense. However a function $f(x) = a\,x^\alpha$ scales for α positive or negative, and α is called the *scaling exponent*. A consequence of scaling, or another way of looking at it, is that any portion of the curve behaves as the whole. The earthquakes in the whole Richter scale 1 to 10 have the same distribution as those with magnitude between 6 and 7; so in this range the earthquakes of magnitude 6 are much more frequent than those of magnitude 7. Limiting Pareto's income distribution to a subset of the rich we find that most of the wealthy people own much less than the super-wealthy, etc.[5]

A concept related to scaling is *self-similarity*, that is, at any degree of magnification a portion of the curve is similar to the whole. Many phenomena in the real world show some degree of self-similarity, which is perfectly met if their mathematical description is a power law. In particular self-similarity is a property of *fractals*, mathematical objects that can be represented as geometric shapes that can be divided into parts, each of which is a smaller reproduction of the whole. Most people will have heard of fractals. Without getting into a complex mathematical analysis we recall that, unlike common objects in Euclidean geometry that have an integer dimension D ($D = 1$ for lines, $D = 2$ for surfaces, $D = 3$ for solids, etc.), fractals have intermediate non integer dimensions.

Many natural phenomena seem to have a fractal structure, at least approximately. For example the vascular networks that distribute resources within an organism, like the blood circulation system, have a dimension D between 2 and 3. Unlike surfaces ($D = 2$), if these fractals are observed down to a finer and finer scale they appear to fill most (but not all) of the space. According to some leading biologists this explains some controversial observations on living organisms, although their theories need further validation. And here power laws come in again.

It is well known that the larger an animal, the lower the metabolism B (essentially the amount of food consumed) compared with the animal's mass M. For a long time the amount of heat H exchanged with the outside of the animal's body was considered a measure of metabolism, that is $B \propto H$. Now heat exchange takes place through the animal's skin, that is a two-dimensional surface in the linear size L of the animal. We then have: $B \propto L^2$, while $M \propto L^3$; therefore $B \propto M^{2/3}$. A power-law, but not the correct one.

In fact, in the early 1930s Max Kleiber made a famous series of measurements from which he concluded that B and M are indeed related by a power law, but the scaling exponent is $3/4$. That is:

$$B \propto M^{3/4} \quad \text{(Kleiber's law)}. \tag{7.6}$$

This surprising result has been justified in the 1990s considering that the

---

[5]Most people do not consider this a particularly painful truth. Apparently the function "pain" does not scale.

capillary network for distributing nutrients, that is the organ responsible for metabolism, has in fact a fractal structure of non-integer dimension $D > 2$, finally yielding the 3/4 exponent of M for measuring metabolism and other biological parameters.

Power laws, then, are encountered everywhere when modeling networks in mathematical terms. In particular they show up in large systems guided by self-regulation rules based on local knowledge, without the intervention of a central control: a nice family of anarchic networks that include the Internet and the Web. No wonder, on the basis of what we have seen in Chapter 6, that a combination of randomness and preferential linking is a crucial ingredient for their evolution.

7.4   The DNS tree: between names and addresses

It is now time to discuss how the concepts seen thus far apply to the topic with which we are concerned most: namely the functioning of the Internet, seen both as a physical network of processors and as a means to organize and retrieve data. To this end we start from a tree that we have already encountered.

In Chapter 3 we have seen that the devices connected to the Internet have an *IP address* through which they can be reached. Devices also have *domain names* organized in a hierarchical structure called *DNS* (*Domain Name System*) *tree* (see Figure 3.9). Roughly speaking, a domain name indicates what we are looking for and the corresponding IP address indicates where it is located. The DNS tree, a basic structure for mapping names to addresses, is replicated in many computers called *name servers*. Any device looking for a resource in the network only needs to know the name of the resource and the address of a name server where the needed information can be found.

The node at the root of the DNS tree, labeled with a silent dot, is the most important of all and corresponds to ICANN, the *Internet Corporation for Assigned Names and Numbers*. The children of the root have a *top level domain* (TLD) name, for example .com, .edu, .gov, etc., to denote particular categories of users; or .us for the U.S.A., .cn for the People's Republic of China, recently .eu for Europe etc., to denote different countries. Going down the tree, each successive level inherits the name of the parent with an additional word to the left of the label, with all these words separated by a dot. For example www.di.unipi.it is the name of the Web server of the university department where two of the authors of this text work, where .it denotes Italy, .unipi denotes the University of Pisa, and .di denotes the Department of Informatics. The name www.di.unipi.it is located in a leaf of the tree from where the corresponding IP address 131.114.3.18 is found. Before proceeding, a few things about ICANN, names, and addresses should be considered.

When the network started operating in the 1970s, a unique archive of domains and addresses was maintained at the Stanford Research Institute. Other users would copy it at night. With network growth the archive became huge and the centralized repository was abandoned in favor of the present distributed solution. The structure of domains is strictly hierarchical. Each node of the DNS tree is managed by a specific institution called an *authoritative name server* that dictates the rules of operation for the domain, maintains consistency of its children, and can assign authoritative name servers to them. Conversion tables between names and addresses are distributed and highly replicated between different nodes so that an address request travels through the network only to the closest location that can answer the query.

In principle a DNS node has full control over its domain and can even cut the Internet connection of a child if some rules are not observed. So ICANN has full control over the whole name system. As ICANN reports to the Department of Commerce of the U.S.A., the government of this country has control over the Internet, at least in theory. In the mid 2000s there was extensive international debate on this issue, with several countries asking for the control of ICANN to be passed over to the UN. Whether this would be a good move is difficult to say. In practice ICANN is directed by an international body that has always acted with impartiality, and, at the end of the decade, the discussion on the controller of ICANN seems to have lost most of its interest. Another point, however, is in full evolution.

Up until the year 2008 an *IP* address, like the 133.114.3.18 mentioned above, was a 32-bit number as required by IPv4, the *Internet Protocol version 4* released in 1981. Thus $2^{32} \sim 4.3 \times 10^9$ addresses could be specified, although due to problems of address distribution no more than one half of the possible numbers were actually available. But even 4.3 billion addresses are insufficient to satisfy the ever increasing world of Internet-connected PCs, smart phones, consoles, and a wealth of other gadgets of the future. In particular Europe and China are running short of numbers and a complex system of private addresses flourishes inside large organizations, to be translated in real time and funneled to the public single IPv4 addresses of the Internet. Rather than just a single computer, an IP address denotes an access point to the Internet through which a whole private network can be connected.

There has been a clear need for a new protocol for a long time. After the unused experimental version IPv5, the long term solution was given in the new protocol IPv6, released in 1995 but not used until recently. Among many innovations, particularly in the authentication and encryption of data streams, IPv6 addresses are specified as 128-bit numbers, so a huge world of $2^{128} \sim 3.4 \times 10^{31}$ different objects can be targeted. Of course adopting the new protocol implies making investments in new equipment, so for many years network service providers have resisted the move to IPv6, especially as the two protocols are natively incompatible. A good opportunity for moving on was offered by the 2008 Olympic Games in Beijing, when the People's Republic of China decided to broadcast all the events on IP television under IPv6. At the end of the decade both protocols are in use, though IPv4 is used much more widely than IPv6. The major operating systems now support both protocols, although their co-existence causes new problems of address translation and communication security. No doubt IPv6 will definitely prevail, but it is hard to foresee when this will happen.

Yet another major innovation was announced in 2008. As we have discussed, the top level domains of the DNS tree had fixed extensions established by ICANN. There are about twenty generic extensions like .com for commercial use, .gov for US government institutions, .edu for colleges and universities, etc.; and about two-hundred and fifty country code extensions like .us and so on. For a long time a community of *cybersquatters* engaged in the unpleasant practice of registering all domain names that could have been interesting for a particular business. Whoever wanted to use one of these names had to buy it back from the assignee, of course at a much higher price than the one required for the original registration.[6] Partly to overcome this phenomenon, and partly as a source of extra revenue, in 2008 ICANN announced the possibility of extending the top level layer of the tree by allowing users to invent their own extensions at a substantial cost. As owner of the extension, an organization could set up its own site without the possibility of intrusions.

7.5 The Internet graph

Being composed of computers and connections among them, it should be possible quite naturally to represent the Internet in the form of a graph. As we shall see, however, the situation is not that obvious because the present structure of the network is a little complicated.

Up until the early 1960s, the worldwide communication was dominated by the telephone network. The basic technique was known as *circuit switching*, implying that a connection between two users was established through a circuit dedicated to them for the whole duration of the call: a reasonable approach since a telephone conversation consists of a continuous stream of signals with a propagation delay negligible for humans. The first attempts to build computer networks were based on similar methods, but in fact something completely different was needed.

During the early 1960s several independent studies emerged on a new way of linking computers rather than humans, where the information does not need to flow continuously on a line as long as it can be reconstructed correctly by

---

[6]As everyody knows, Italians are very fond of "pasta" (noodles). When e-commerce started in the country many names having "spaghetti," "fusilli," "lingiuine," etc. in them were readily registered, with the sole aim of selling them back to pasta producers. Even the Italian top level domain .it has been exploited because of its English meaning, giving rise to sites like buy.it, lift.it, use.it, rent.it, cook.it....

the receiver. The technique of *packet switching* then emerged, marking what is probably the biggest difference between telephone and Internet communications. According to this technique, data are disassembled into packets to be sent independently from one another and reassembled upon arrival, and the path connecting two points can change dynamically depending on momentary link availability. In the minds of its inventors this technique had several advantages, although it required a tough probabilistic approach on queueing and sophisticated routing algorithms to ensure the correct delivery of all messages. One is that a link could be shared by different messages hosting their packets intermingled, making better use of the network. Another is that a message, or rather some of its packets, could be de-routed through a different path in case of traffic jams or hardware crashes. And, perhaps most importantly of all, the network could continue working even in case of destruction of some of its nodes.

In fact, in October 1962 the so called Cuban missile crisis had brought the world very close to nuclear war, and the Defense Advanced Research Projects Agency (DARPA) of the United States, together with M.I.T., had started studying the connection of strategic centers with a technology that could resist enemy attacks. This was the seed of the early Internet, initially called ARPANET. The network, then, was born for a military interest, although the world has to credit the U.S. Department of Defense for having left the scientists absolutely free to act in a manner that has brought benefit to all.

By mid-1968 the ARPANET project was complete. Contemporary routers were small computers called IMPs (*Interface Message Processors*) in charge of packet switching. IMPs were connected to external leased lines, and each host computer was in turn connected to an IMP. The initial configuration of the network had four nodes located at U.C.L.A., U.C. Santa Barbara, the University of Utah, and the Stanford Research Institute. History records that the first message, consisting of the word "login," was transmitted from U.C.L.A. to the Stanford Research Institute on October 29, 1969 and crashed at the third letter. One hour later the problem was solved and the whole message got through.

From the early structure of the network to the present Internet topology many significant steps have been taken. Aside from the obvious difference in size between the ARPANET of the early 1970s and the present Internet, that makes the two networks quite incomparable, the major change occurred when the number of participants and the global set of connections was released from central control. The Internet then became a structure growing and changing in countless locations at unpredictable times according to local needs, as long as the basic rules of the game were observed. Its present structure was not designed by anyone. In order to enter into the system, one only needs to obtain an IP address from an authoritative name server and use the standard connection protocol TCP/IP. In the words of Tanenbaum, one of the major world authorities in computer networks:[7]

... a machine is on the Internet if it runs TCP/IP protocol stack, has an IP address, and can send IP packets to all the other machines of the Internet.

The protocol TCP/IP (*Transmission Control Protocol / Internet Protocol*), strictly related to packet switching, became the network standard as early as 1983. We will not enter into details that have to do with the structure of system software.[8] It suffices to recall that the two sections of the protocol essentially serve two different requirements for packet transmission. IP (in either the IPv4 or the IPv6 version) dictates the rules for sending each single packet along the network from one node to another, on the basis of the final destination whose IP address is specified in the packet itself. The route actually followed by a packet is decided locally at each hop from a node to the next, depending on a presumed shortest path to the destination, and on the traffic situation. As a result packets are finally received at different and generally unpredictable times, and may even arrive in a different order from the one in which were sent if they have followed different routes because of traffic problems. The role of the TCP layer is to reconstruct the message, placing the packets in the right order.

Another basic innovation of the Internet was related to the possibility of putting together different networks that were created independently. The topology of the network grows by adding new subnetworks at each step, since each newcomer generally gets in with much more than a single computer. Quoting Tanenbaum again:

The Internet is not a network at all, but a vast collection of different networks that use certain common protocols and provide certain common services.

Each single network participating into the game is called an *autonomous system* (or AS). It may be an end user like any organization connected to the Internet, or an *Internet service provider* (ISP) which takes care of dispatching traffic to its customers. The situation is quite complicated: by and large the graph describing the Internet has the structure of Figure 7.4.

ASs are connected to the Internet through their *routers* dedicated to traffic dispatching. Any AS may contain all sorts of machines serving the organization, connected in a network whose structure and internal communication rules are decided by the local administrator. The routers, however, must communicate with the routers of the other ASs using a common protocol. Today's standard is *BGP* for *Border Gateway Protocol*.[9] According to the Internet

---

[7]Andrew S. Tanenbaum, *Computer Networks*, see bibliographic notes.

[8]The field of network protocols is densely populated with acronyms of two, three, or four letters. It is an unbearable world of insiders from which any person of good taste is tempted to escape. We will stay away from it as much as possible.

[9]BGP makes use of TCP/IP. In technical terms BGP is in a "higher" application layer; TCP is in the transport layer; IP is in the network layer.

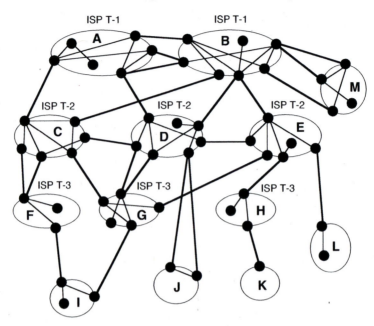

FIGURE 7.4: A portion of the Internet graph. Autonomous systems are shown in ellipses: among them, A to H are ISPs distributed in the tiers (T) 1 to 3. Black dots are routers: some of them are connected only internally to the AS for dispatching local traffic. ASs I to M are networks of external users purchasing Internet access from providers of different tiers.

philosophy, BGP has the role of routing messages as fast as possible without applying any control on the flux of data. A purpose that contains a good deal of hypocrisy, particularly if the AS is an Internet service provider, because the protocol allows some manual reconfigurations at the router level, e.g., blocking the transmission of unwanted messages.[10]

Internet service providers, in turn, are classified into three *tiers* (see Figure 7.4). An end user pays an ISP for Internet access. The ISP in turn may pay an 'upstream' ISP for reaching ASs that the former ISP cannot reach, up to an ISP of Tier 1 that can connect to any point of the Internet without paying in turn for the service. Several Tier 1 ISPs are huge communication companies like AT&T or NTT. They reach everywhere by direct connections of their routers or by *peering*, i.e., they traverse the other ISPs of Tier 1 free of charge. This commercial policy seems to be peculiar to the Internet, where some big business competitors find it beneficial to exchange services on a settlement free

---

[10]The most impressive control on Internet communication takes place in the People's Republic of China where all traffic is routed through some nodes of the national network where harsh censorship is exerted. See the bibliographic notes.

basis (obviously at the expense of the smaller guys who pay for accessing their tier). Tier 2 ISPs are generally (but not always) smaller. They peer with other networks at the same tier but have still to pay an ISP of Tier 1 to reach some points of the Internet. Tier 3 ISPs are even smaller. They lay at the periphery of the network and always purchase Internet access. Because it is generally believed that the providers of Tier 1 provide better service, important ASs tend to connect with them directly.

Figure 7.4 shows that the Internet can be seen as a set of routers clustered into subnetworks. The huge number of other machines internal to each autonomous system, although connected to the Internet through their routers, do not take part in the global traffic flow and do not contribute to the dynamic evolution of the network. It is then customary to study the Internet graph at *router level*, with all the connections among them: this is the graph of black dots and connecting lines of Figure 7.4; or at *AS level*, i.e., considering each AS as a node and taking the connection between two ASs as a single arc even though they may share many physical links: this is the graph shown in Figure 7.5, and here the mathematical laws of graph growth come in.

The first observation is about the size of the two graphs. Of course a precise answer cannot be given because the network changes (and essentially grows) continuously. But even a trustworthy approximation is impossible because the network evolves without a central control. It is true that, due to certain protocol technicalities, autonomous systems may be identified by a 16-bit number assigned centrally, so a maximum of 65,536 ASs can be cataloged this way. But a vast majority of ASs do not have an official number, so we can only try to suggest an order of magnitude of some hundreds of thousands for the year 2010. There may be one hundred times more routers. The statistical properties of the two graphs have however been observed in countless experiments and are well known.[11]

First of all both graphs are small worlds. Although a direct evaluation of the mean shortest-path length $\bar{l}$ as a function of the total number of nodes cannot be performed because the latter number is unknown, experimental observations allow us to conclude with reasonable confidence that $\bar{l} \simeq 4$ for the AS graph, and $\bar{l} \simeq 10$ for the router graph. This is combined with high clustering. In the AS graph the mean value of the clustering coefficient is $\bar{C} \simeq 0.2$, much greater than that of a random graph. The value of $\bar{C}$ in the router graph is even higher, but it is not particularly significant because it is strongly influenced by geographical factors as close routers have a much higher probability of being directly connected than those separated by a greater distance.

Let us now examine a major indicator of network growth, namely, the node degree distribution $P(d)$ for the AS and for the router graphs. A great amount of experimental data has been collected in recent years (see the bibliographical

---

[11]Up until 2006 raw data on the Internet were collected by the National Laboratory for Applied Network Research (NLANR). Since then different institutions have begun making similar collections.

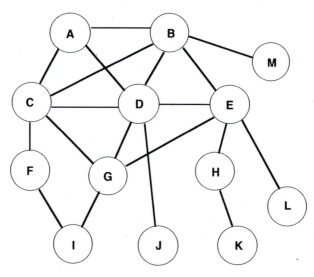

FIGURE 7.5: The portion of the Internet graph of Figure 7.4 reduced to the AS level.

notes). All experiments agree on a power law for the function, that is $P(d) \propto d^{-\gamma}$, generally with $\gamma \simeq 2.2$ for ASs and $\gamma \simeq 2.3$ for routers. However we are more interested in the behavior of the curves than in the absolute value of the exponents. Evidence has also been given that the basic aggregation rule is preferential (i.e., new nodes are attached with higher probability to existing nodes of higher degree), although a certain amount of perfect randomness, and some geographical influences, also play a role. According to what we have seen in Section 6.3 of the previous chapter, the overall power law behavior is consistent with the growth mechanism.[12]

The experimental plots of $P(d)$ show some peculiarities that must be carefully considered. Starting with the degree distribution at AS level, a curve in a Cartesian plane with coordinates $d, P(d)$ cannot practically be drawn because of the enormous difference in the degrees among different nodes. Many ASs have very few connections with the Internet, possibly as few as one or two, while the ISPs of Tier 1 may have several thousand. The power law distribution has a fat tail and in fact the ASs "rich" in connections are many more than for a random exponential decay. But this tail is also "long," meaning that many nodes laying towards the right end of the curve have very high degrees that are generally different from one another. As a limit the degrees of the "richest" ASs occur just once (i.e., with the same probability $1/n$). For

---

[12] Additional mathematical models have also been proposed for explaining some features of Internet dynamics, see the bibliographical nodes for further investigation. All these models are consistent with an overall power law distribution of $P(d)$.

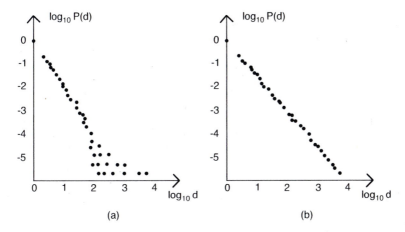

FIGURE 7.6: (a) Degree distribution $P(d)$ of the Internet graph at AS level, in logarithmic scale. (b) Cumulative distribution $P_c(d)$ for the same data. The two curves do not represent a specific experiment, rather they have the same behavior as most of the experimental curves reported in the literature.

example several Tier 1 ISPs have degree around three thousand and these degrees are all different. As a consequence $P(d)$ can be practically represented only in logarithmic scale for coping with the great difference of degree values, see Figure 7.6(a). The interpretation of the diagram should be immediate: for example for $\log_{10} d = 1$ (i.e., for $d = 10$), we have $\log_{10} P(d) \simeq -2$ (i.e., $P(d) \simeq 1/100$); that is, about one over one hundred ASs have ten external connections. Up to a certain value of d the curve can be neatly interpolated with a straight line whose slope equals the exponent γ of the corresponding power law, however, the tail of the distribution is quite messy as shown in the figure.

As anticipated in Section 6.3 of the previous chapter, we get a satisfactory representation of the whole curve by resorting to the cumulative distribution $P_c(d)$, that is, to the probability of finding a node of degree $\geq d$ instead of exactly d. The new curve is shown in Figure 7.6(b). Note that each value of d corresponds to a unique value of $P_c(d)$. Being the integral of $P(d)$, the new distribution follows a power law with exponent $\gamma_c = \gamma + 1$.

Curves similar to those of Figure 7.6 have been found also for the graph at router level, although the information here is less certain because the connections internal to each AS may be unknown. As already discussed the value of the exponent γ (hence γ_c) for this graph has been generally reported to be higher than that of the AS graph and the approximating straight line is steeper. On average routers tend to have fewer connections than AS nodes.

A final parameter that we have not yet considered characterizes the importance of Internet nodes from a traffic point of view. Introduced in the realm of social sciences, the *betweenness* $\beta(w)$ of a node w indicates the centrality of w with respect to a network where all communications between each pair of nodes u, v follows the shortest path between them. For a directed or undirected graph, betweenness is formally defined as:

$$\beta(w) = \sum_{u \neq v} \frac{S_w(u,v)}{S(u,v)} \qquad (7.7)$$

where $S(u, v)$ denotes the number of shortest paths between u and v; $S_w(u, v)$ is the number of such paths that pass through w; and the sum is performed over all pairs of connected vertices (i.e., for which $S(u, v) \neq 0$). In the graph of Figure 7.5 there are three shortest paths between nodes I and B, in fact, I-F-C-B, I-G-D-B, and I-G-E-B, and one of them passes through node D. Therefore we have $S(I, B) = 3$ and $S_D(I, B) = 1$, and the contribution of nodes I, B to the betweenness of D is $1/3$. As we might expect, nodes with high degree tend to have high betweenness. Some experiments have demonstrated that the distribution of betweenness in the AS graph also follows a power law and that nodes of high degree tend to have high betweenness.

Clearly the value of betweenness shows the importance of each node for traffic purposes if shortest paths are used in communication, as is preferred by the Internet protocols as long as the links along the path are not too busy. Therefore deliberately "attacking" the nodes with high degree and high betweenness may jeopardize the network. In the Internet graph at AS level such nodes are likely to be important ISPs, and these are difficult to destroy because they have many routers in diverse and well separated locations. Attacking the network at the router level is potentially more harmful if the degree and possibly the betweenness of single routers are known. While the readers interested in going deeper into this subject may refer to the bibliographical notes, we recall here a result that can be formally proved. Comparing random networks to those ruled by power laws, the former are much more resistant to hostile attacks, while the latter are more robust against random failures (like, for example, the temporary unavailability of a router in the Internet).

Finally note that nodes with high degree and high betweenness are natural sites for (computer) virus spread, which the Internet is particularly vulnerable to. Like all self-organizing networks, the Internet is naturally well equipped against natural hazards, but malicious attackers may put it in serious danger.

7.6   The Web graph

We have represented the Internet as a graph with autonomous systems or routers at the nodes to show the web of connections between sites, disregarding the huge number of other computers and storage devices that are connected to the network for performing other services, including our own terminals.

As we all know, a common way that information exchanged through the network is organized is in the World Wide Web (or simply the Web), a system of interlinked documents that may actually reside in any computer or storage device connected to the Internet, often replicated in many copies spread around the network for faster access. While in the next chapter we will explain the mathematical tools that allow us to "browse" and "search" through these documents, we recall here how and why the Web was conceived, how it is represented in graph form, and some basic properties of this graph.

In its early years the Internet was very difficult for non specialists to use. A drastic change took place when a system originally developed for coordinating scientific projects at CERN, the major European center for nuclear physics, was made available over the Internet in the early 1990s. The Web was born. The intent of the inventor, the British computer scientist Tim Berners-Lee, was to allow independent actors to conduct their own transactions over the network without a centralized control. The well known concept of hypertext, which we will discuss again later, was adopted for this purpose. All documents are represented in a common format using a special language, HTML (Hyper-Text Markup Language) or its extension XML, that supports links to other documents. These can be reached by clicking on hot spots.

The integration of hypertext with the Internet had spectacular consequences. Web *pages*, rendered attractively on a computer monitor, allowed one to jump from one another following the links and to come back, through a unified system of identifiers called URLs (Uniform Resource Locators). Since links are unidirectional one can jump to another page without permission, as long as the page is on the Web. In principle it is possible to develop entire Web servers without permission.

It is natural, then, to represent the Web as a directed graph with pages as nodes and links as arcs. This graph is somehow "hosted" by the Internet graph, but has no resemblance to it. Figure 7.7 repeats the four autonomous systems K, H, E, L of Figure 7.4, with the addition of three new computers (encircled) in K and L connected to the local servers. The three rectangles are Web pages hosted in these computers. Pages 1 and 2 are connected by a directed arc (dashed arrow), however, they reside in two computers of the same autonomous system and are then invisible to the Internet graph both at AS and router level. Page 3 and 2 are also connected by an arc, but their physical connection in the Internet graph passes through several ASs or routers.

Over the years many articles have been published on the size of the Web graph. Search engines declare the number of pages that they can reach thus establishing a lower bound on the Web size.[13] In a blog written by Google engineers a total of one trillion pages reached up to 2008 was claimed, but a definitive answer is not obvious as many (possibly most) pages belong to the so called *deep Web* that is not directly addressed by the search engines.

---

[13]The sets of pages reached by different engines overlap partially, but the size of their union is not known. Strictly speaking the lower bound is given by the greatest of the numbers declared.

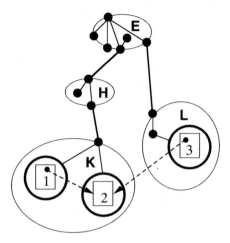

FIGURE 7.7: Web pages 1, 2, 3 and their links.

While this point will be made clear in Chapter 9, it is sufficient to say now that the Web graph is by far the largest discrete mathematical entity humans have ever directed their attention to.

From what we have seen in Section 7.1, it is not surprising that the Web graph contains a giant component like the one in Figure 7.2. Several experiments show that this is indeed the case, with the subgraphs GI, GS, and GO containing a vast majority of the nodes, and with the other nodes shared among tendrils and small independent components. Recent studies suggest that about two thirds of all Web pages belong to GS. What is maybe even more interesting is that the GS topology scales, that is, the subgraphs of GS tend also to have the structure of Figure 7.2.

Furthermore it is generally said that the Web graph, or better its giant component, is a small world, although such a statement must be taken with some caution. It has been observed that, if all the arcs were not directed, the mean distance between any two nodes would be about six, the magic number of Milgram. But this does not mean much. The mean distance in the actual (directed) graph has been reported of being approximately sixteen at the beginning of the decade 2000, and is slowly growing with the ever increasing number of nodes.

The degree distribution of the Web graph is separately studied for the in-degree and the out-degree that in principle are independent from one another. In fact the outgoing links of each node are determined by the needs of the page owner and are rarely changed after the page is created, while the in-degree is completely out of the page owner's control.[14] All experiments have shown that

---

[14] A certain dependence between in and out degrees is due to the presence of "mutual references." That is, two pages may be built deliberately pointing to each other.

both the in-degree and the out-degree of the Web nodes exhibit a power law distribution with exponent γ between 2 and 3. Of the two degrees the former is much more interesting since the number of incoming links is a measure of node "popularity." To better understand all this let us consider empirically how the Web grows: in particular how a new page appears and is linked to the existing Web.

Unless the author of the new page is particularly expert, for example a professional designer of Web sites, more than likely he or she will get inspiration from an existing page. This is easily done thanks to the availability of the source code of most Web pages, that is explicitly made public: even some outgoing links may be copied if the inspiring and the inspired pages deal with similar business (see below). In any case the major attachment process is preferential, with important or popular pages being pointed to with higher probability. Note also that, for the new page to be found by search engines, at least one link to it must be provided either from the site of the organization to which the page owner belongs, or from some other source, for example a blog.

Several Web growing models have been proposed in mathematical terms, built on the scheme of the preferential and random process 4 presented in Section 6.3 of the previous chapter, extended to take care of some real-world features like adding and possibly deleting several links at each step, with different probabilities.

On practical grounds a drawback of most preferential linking models is that the choice of links requires the knowledge of the degrees of all nodes of the graph to establish preference. Partly for this reason an interesting *copying model* has been proposed, aimed at capturing the tendency to imitate existing pages. Figure 7.8 shows the difference between preferential linking and copying.

All models, including the copying one, lead to a power law for the vertex in-degree and out-degree distributions, with exponents close to the experimental values already mentioned. This does not imply that such models capture all the features of the Web evolution, but at least shows that they are not unreasonable. The mathematical analysis is quite complicated, so we refer to the bibliographical notes for that. It is worth noting, however, that the fat tail for the in-degree power law is particularly scattered because there are many popular pages with very high and different in-degrees (the Yahoo! home page with hundreds of thousands incoming links is often mentioned). So the cumulative distribution must be plotted for clarity, as done for example in Figure 7.6 for the Internet graph.[15]

Unlike in the Internet graph, betweenness is not a particularly important parameter for Web search because users tend to extend search paths only for a few links, relying more often on the direct answers of a search engine than

---

[15]This point is often not sufficiently underlined in the literature.

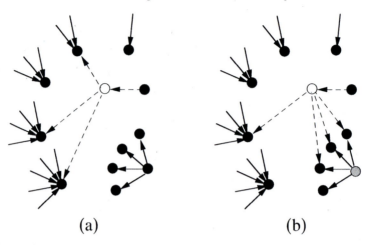

(a)   (b)

FIGURE 7.8: A new page appears in the Web as a white node. (a) Effect of preferential linking. (b) Effect of copying the grey node: the new page copies most of the links of the copied page.

on the possible successive hops through a chain of links. Instead, betweenness is related to the presence of particular subgraphs called communities.

7.7   Graph communities and the Web

Loosely speaking a *community* in a graph G is a subgraph C whose nodes share only a few arcs with the rest of G and are densely connected inside C. In fact there is no unique definition of community notwithstanding the great importance of this concept in social sciences and in biology, where a community is a group of individuals sharing the same interests or characters; and in the Web, as we shall see.

No matter how a community C is defined, its detection is a hard problem as C must be selected from among a huge, often exponential, number of possible subgraphs. For almost all the definitions of community that have been proposed the detection problem is NP-hard, a term explained in Chapter 4, so only heuristic algorithms can be used possibly missing some relevant solutions. An oversimplified example was given in the detective story of that chapter (Section 4.5), where a criminal group was defined as a clique in the Web graph. Finding a clique is a well known exponential problem.

In the realm of networks the definitions of community are not as stringent as that of a clique, but still exact detection is exponentially difficult. A rea-

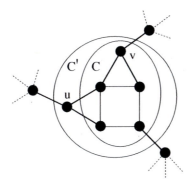

FIGURE 7.9: An α-community C (internal oval): each node in C has more connections inside than outside C. Note that also C' (external oval) is an α-community.

sonable definition coming from graph theory and applicable to a directed or undirected graph G is the following:

An α-community is a subgraph C such that each node of C has more arcs pointing to vertices inside C than to vertices outside C. (7.8)

The simple example of Figure 7.9 shows an α-community C in an undirected graph, and enlightens a major problem that unfortunately arises with most definitions of communities. Not only is C an α-community, but also C' that contains C is an α-community. And more seriously, the subgraph obtained from C excluding node v and including u is also an α-community. That is, communities can totally or partially overlap. This is why finding them is so difficult, and why their relevance in real life problems is sometimes questionable. In any case, something can be done.

A well known method applied in social sciences, due to Mark Newman and Michelle Girvan, uses the concept of betweenness to define and detect a hierarchy of communities that are contained inside one another. Betweenness is now computed on the arcs instead of the nodes, with an immediate extension of relation (7.7). Cutting the arcs in the order of decreasing betweenness eventually divides the graph into consecutive halves that exchange a "large flow of information" between them. Since such a large flow is supposed to traverse the links between social communities, these are recursively identified with the two halves. For this and for many other heuristics that have been proposed, we address the reader to the bibliographical notes. It is also worth noting that the problem could be treated in a distributed fashion, an approach that we will examine in the next chapter. However, no major attempt has been done in this direction.

Let us now focus on the Web graph. A Web community is a set of pages with some strong property in common and, as in all other fields, communities are often tightly connected internally and have a much lower density of connections with the rest of the graph. Social networks form communities in an obvious way. Large service providers and companies engaged in e.commerce routinely detect communities of customers with similar needs or requests.

Other kinds of Web communities are also important, such as *bipartite* subgraphs that are somehow related to Web page retrieval as we will explain in Chapter 9. These subgraphs are formed by two subsets A, B of pages where each page of A points to all the pages of B but does not point to the other pages of A, while no assumption is made on the arcs leaving from B. Such subgraphs are often taken as indicative of a family A of actors with some interest in common displayed in the pages of B, but in some sort of a competition between each other. For example the pages of A may belong to car dealers selling the same car models that are described in the pages B of the manufacturers. Other communities, known as *spam farms*, are fraudulently built to attain undue advantages from other users, as will be explained in Chapter 9 after having discussed how search engines work. The algorithmic game of detecting all these communities is fully open.

Bibliographic notes

The topics of this chapter are treated in a vast literature, some already mentioned in the bibliography of Chapter 6.

Watts and Strogatz presented their small-world model in the article: Watts, D.J. and S.H. Strogatz. 1998. Collective dynamics of "small-world" networks. *Nature* N. 393, pp. 440-442. Barabási and Albert originally discussed the preferential growth of networks in: Barabási, A.L. and R. Albert. 1999. Emergence of Scaling in Random Networks. *Science* N. 286, pp. 509-512. Kleinberg presented his results on small world routing at the highly reputed ACM Symposium on Theory of Computing in 2000. A resume of his work was published in: Kleinberg, J.M. 2000. Navigation in a small world. *Nature* N. 406, 845.

Then, two important books appeared aimed at general public: Barabsi, A.L. 2002. *Linked: The New Science of Networks*. Perseus Publishing, Cambridge, MA. Then: Watts, D.J. 2003. *The Science of a Connected Age*. W.W. Norton & Co., New York. As a more technical reference on the theory of networks let us mention among others: Bornholdt, S. and H.G. Schuster, Eds. 2002. *Handbook of Graphs and Networks*. Wiley–VCH, Berlin. These books, however, deal with networks in general and make almost no specific reference to the Internet.

It is worth noting that there has been some confusion on scaling and self-similarity, as in the literature on "scale-free" networks these concepts are often left at an intuitive level. A rigorous discussion can be found in: Li, L. *et al.* 2005. Towards a Theory of Scale-Free Graphs. *Internet Mathematics* Vol. 2, N.4, pp 431-523. One of the most complete and readable surveys on graph communities is: Fortunato, S. and C. Castellano, 2009. Community Structures in Graphs. In *Encyclopedia of Complexity and Systems Science* (R. Meyers, ed.). Springer p. 1141.

A reference book on the organization of the Internet and its protocols is: Tanenbaum, A.S. 2003. *Computer Networks*. IV Edition. Pearson Education. To find out more about Internet censorship, a very good reference is still an article of some years ago: Cherry, S. 2005. The Net Effect. *EEE Spectrum* Vol. 42 - 6. IEEE Press.

For the Web graph, an early well known study was conducted by a group of computer scientists from different institutions and is reported in: Broder, A. *et al.* 2000. Graph structure in the Web. *Networks* N. 33, pp. 309-320. A recent interesting presentation in mathematical terms is: Bonato, A. 2008. *A Course on the Web Graph*. American Mathematical Society, *Graduate Studies in Mathematics*, Vol. 89. In particular the book contains a survey of different mathematical models for the Web graph.

An important study of the role of fractals in biological systems, that includes a proof of Kleiber law, is contained in the article: West, G.B., J.H. Brown, and B.J. Enquist. 1999. The Fourth Dimension of Life: Fractal Geometry and Allometric Scaling of Organisms. *Science* Vol. 284, pp. 1677-1679. The reader should be advised that the authors' approach has raised several criticisms, and their theory is still under discussion.

A final reference comes from the general literature. Alessandro Manzoni's novel *The betrothed* (*I promessi sposi*) appeared in an excellent English translation in The Harvard Classics, edited by C.W. Eliot, P.F. Collier & Son, New York 1909–1914. Nowadays it can be found at: http://www.bartleby.com/21/: our citation in English is taken from there. Of course this reading is recommended only to those interested in 19th century literature and has no direct relation to the Internet.

Chapter 8

Parallel and distributed computation

How several autonomous computers can communicate and cooperate to solve problems too large for a single computer.[1]

The potential power of multiple autonomous computers connected together in a network is raising high expectations in an ever increasing number of people. What was familiar before only to a group of specialists started to become clear to anybody with the advent of the first web navigator programs: something really new was happening, as is well expressed in the words of a recognized *guru* of popular science:

In the years roughly coincidental with the Netscape IPO, humans began animating inert objects with tiny slivers of intelligence, connecting them into a global field, and linking their own minds into a single thing. This will be recognized as the largest, most complex, and most surprising event on the planet.[2]

Other events and other equally revolutionary discoveries have raised similarly enthusiastic comments and reactions in the past. What will be remembered of our planet we simply do not know. Perhaps nothing at all will be remembered if its systematic destruction continues at the present pace. But if some things do remain, computer networks will be among them.

Rather than tiny slivers of intelligence we will more cautiously speak of *entities* at a much greater scale such as computers, programs, network sites, or even clusters of computers or sites. Each entity is able to execute algorithms with great accuracy and very high speed, independently of any other entity. When connected, these entities are able to agree on a common strategy and cooperate to the solution of a problem, working together as a whole. It has often been remarked that such organization mimics the functioning of neurons

---

[1]A role of this chapter is to show how complicated can be the parallel or distributed solutions of apparently simple problems. Some parts, in particular sections 8.3 and 8.4, are quite technical. Uninterested readers may skip the distributed protocols and other technicalities, but are advised to give some attention to the subtle logical concepts that are at their base.

[2]Kevin Kelly, We are the Web, *Wired*, issue 13-08, August 2005.

inside the human brain, and that this interpretation could be a guideline for the development of artificial intelligence, although substantial progress in this direction is not easy to foresee. More humbly we will explain what may be expected from the cooperation among entities from an algorithmic point of view.

Until now we have assumed that algorithms are executed step by step according to a mode of *sequential computation* which reflects the work of one standard processor in a centralized computer system.[3] The computation becomes *parallel* if a group of processors (entities) cooperate in order to solve a problem by continuously exchanging data and results. The algorithmic strategy must change to exploit the capabilities of the whole system, while the organization of the work becomes critical. Sometimes a parallel computation is called *distributed*, the opposite of centralized, while the term parallel is the opposite of sequential. There is some disagreement about the difference between the two terms parallel and distributed when related to computation, a debate which is perhaps not so interesting at this level of discussion. In both cases there is a set of entities working together on a communication network. By and large we will use the term parallel when the overall computation follows a predefined global strategy, and the term distributed when individual computers operate on a network independently from one another. In both cases the problem to be solved may come from the outside or have to do with the connecting network itself: what is most relevant in our context is that both modes of operation arise in Internet algorithms.

8.1   The basic rules of cooperation

A parallel or distributed computation must obey some general rules. As we know, the same problem can be solved using different algorithms, and the most important measure of efficiency is the comparison of execution time. While discussing sequential computation we have seen that time is conventionally measured in terms of elementary steps instead of seconds to make the reasoning independent of the specific computer used. When going from sequential to parallel or distributed computing a fair comparison can be carried out only by assuming that all the entities taking part in the game are identical, but measuring time in terms of elementary steps is still useful. With this in mind, we start with a preliminary consideration. Let T_{seq} be the time required by the best sequential algorithm known for a given problem. Before thinking of a possible parallel solution, let us try to understand how much the execution

---

[3]Inside a computer many elementary actions are actually performed in parallel, in order to execute a single step of an algorithm that is sequential at a higher level. The latter is the level to which we refer in the text.

time can decrease using $P > 1$ entities. If T_{par} is the time of any hypothetical parallel algorithm, we must have:

$$T_{par} \geq T_{seq}/P \qquad (8.1)$$

where the maximum benefit we can hope to achieve, given by the equality case of the expression, is a uniform assignment of the steps of the sequential algorithm to the P available entities.

Relation (8.1) is easily justified by simple reasoning. Starting from a parallel algorithm A, a single entity could execute all the steps of A sequentially by simulating cyclically the work of the P entities. More precisely the entity should simulate one parallel step of A by executing, one after the other, all the operations made by the P entities in that parallel step, so the simulation of A would require $P \times T_{par}$ total time. If the relation (8.1) were not true we would have: $P \times T_{par} < T_{seq}$, that is the new sequential algorithm would require a time smaller than T_{seq} thus contradicting the hypothesis that T_{seq} is the time required by the best known sequential algorithm.

Taking an example from real life, consider clearing the overnight snow fallen on a driveway. Now, our entities are P guys with shovels. If the entire driveway requires time T_{seq} to be cleared by the fastest shoveler working alone, we cannot expect that the group would complete the work in less than T_{seq}/P, and this can only be achieved if all the shovelers start at the same time, all of them are equally fast, and the time needed to assign the portion to clear to each shoveler is negligible. In particular this last requirement can only be satisfied for a small group. For parallel computers or complex networks we must accept that $T_{par} > T_{seq}/P$ and regard the limit case $T_{par} = T_{seq}/P$ as being very unlikely. Still for a large value of P, that is a parallel system with many entities, the gain in time may be substantial.

Another important consequence of relation (8.1) pertains to problems for which we do not know a sequential algorithm that runs in polynomial time, hence T_{seq} is an exponential function of the data size. Such problems have been defined as *intractable* in chapter 4 and remain intractable even in the framework of parallel computation since, due to relation (8.1), T_{par} remains an exponential function unless the number P of entities is also exponential, and that is obviously unacceptable. In the realm of computational complexity where problems are classified as tractable (polynomial) and intractable (exponential) the computational power of a set of P entities, that is the capability of solving problems of one class or the other, is the same of a single entity unless P itself is exponential.

What does change is the size of solvable problems, although there are cases where a collective approach may be more complex than working sequentially, or may even be impossible if the problem is "inherently sequential." Among countless examples consider what happens in secure communications, where a message is divided into consecutive blocks $m_1, m_2, \ldots$ which are converted into encrypted blocks $c_1, c_2, \ldots$ using a secret key k, and then sent over an insecure

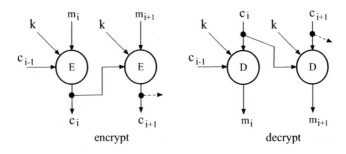

FIGURE 8.1: Scheme of the *CBC* transmission mode. E and D are the encryption and decryption functions, respectively.

line. Since all the blocks are encrypted with the same key the method suffers from a serious drawback, because if two blocks of the message are identical the corresponding encrypted blocks are also identical. For example if $m_i = MILLION\_DOLLARS\_$ and $m_j = MILLION\_DOLLARS\_$, two identical blocks c_i, c_j, will eventually be transmitted, thus leaking some information about the message $(m_i = m_j)$ to an eavesdropper listening on the line.[4]

The problem is adressed in a widely used transmission standard known as *CBC* for *cipher block chaining*, where each block c_i is not only a function of m_i and k as before, but also of the previous encrypted block c_{i-1}, along the lines of Figure 8.1 that shows both the encryption and decryption phases. Two identical blocks of the message generate two encrypted blocks with unpredictable differences for the eavesdropper. It should be clear that the operation of encryption is inherently sequential, since each block c_i can only be computed once the previous block c_{i-1} is known. Decryption however can be performed in parallel with great efficiency once all the encrypted blocks have been received.

In the next chapter we present what is probably the most spectacular success of parallel and distributed processing, namely the functioning (if not the existence) of the Web. It must be also underlined, however, that the possibility of connecting many computers in a network is changing the way computation is developing, because of both the huge power that can be obtained from the concurrent and cooperative use of many processors, and the more efficient usage that can be made of each one of them. Although a firm terminology has not yet been established it is customary to talk of a *grid* as a sort of virtual supercomputer consisting of a large cluster of loosely connected machines that work concurrently to perform a common task. To some extent a grid can perform the same task as a mainframe, but it is much more versatile

---

[4]8 and 16 bytes are among the standard block lengths. $MILLION\_DOLLARS\_$ is coded with 16 bytes.

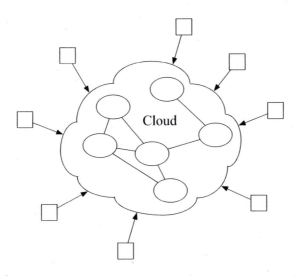

FIGURE 8.2: A cloud infrastructure. Ellipses are sites and squares are users.

because its size is easily adaptable to the problem at hand. Clearly the success of grid computing crucially relies on the inner nature of the problems to be tackled, that is, on the possibility of dividing them into parts to be solved independently before the partial results are recombined.

A grid may belong to a single organization, as done for example by Yahoo! for computing the initial two-quadrillions bits of π using more than one thousand connected computers. More interestingly, a grid may be set up by different organizations contributing their resources to a unique project as in the *TerraGrid*, a huge computing infrastructure whose eleven partner sites are major American Universities and research centers that make their high-performance computers, software tools, and specific databases available through a very fast network accessible for scientific research.

A more recent term related to the concurrent use of computers is the *cloud*, denoting a large infrastructure generally selling computing resources of various kinds over a network. Grids and clouds have many features in common, with the latter term frequently used to refer to infrastructures where external users rent the service under a "pay-as-you-go" contract. Huge clouds have been set up in consortia by giants of the IT market and leading universities. One of these infrastructures is *Open Cirrus* with participating sites in North America, Europe, and Asia, where each site furnishes the cloud with a huge cluster of computers and storage devices, and each authorized user can access any of the sites. In a tradition coming from the telephone industry, cloud infrastructures are indeed represented by a cloud (Figure 8.2).

While clouds are reserved for paying customers or highly skilled researchers as in Open Cirrus, other interesting experiments on distributed computing can be joined on a personal basis as long as a would-be participant is connected to the Internet. In particular, in 2002 the non-commercial *BOINC* software (for Berkeley Open Infrastructure for Network Computing) was released, to be freely entered by anyone wishing to pursue a scientific project where distributed computing can help, tapping into other participants computers through the network for using their currently unused computing power. Volunteers then contribute to the development of science and technology and have the chance of making their own discoveries.[5]

We will now consider the benefits of cooperative working in a strictly algorithmic sense, starting from the realm of parallel algorithms and passing later to the more complex distributed world, with a warning that what is coming is not going to be so easy. The main reason for this is that the theory of algorithms in a parallel or distributed setting is not yet as well developed as for sequential computation and several new complications may be encountered. New problems will arise, sometimes difficult to solve, and sometimes unexpected. Furthermore when several connected entities work together it is generally quite difficult to understand what is going on even if the communication protocol is known in detail, in the same way that an animated conversation is difficult to follow if everybody talks at the same time. Probably the description methods available today are somehow inadequate because it is difficult to explain the parallel behavior of different entities using a traditional *sequential* text such as that of this book.

8.2   Working in parallel: some logical problems

As we will see in the next chapter, search engines constitute a successful application of parallel computation because they crawl the Web to collect pages, build extremely large dictionaries, and look up words in them using vast numbers of computers that work together. This is why such engines are able to answer our queries almost instantaneously. However, serious and unexpected difficulties may arise when several entities work concurrently.

A typical case is encountered when different entities have access to some common resources to be used in *exclusive mode*, thereby blocking temporarily the access of other entities to the same resources. This may generate crossing blocks where entities might get stuck forever when competing for the same resources, as in a famous paradigm of computer science known as *the dining philosophers problem*. Here five philosophers sit at a circular table with

---

[5]A remarkable application of BOINC is the *SETI@home* project aimed at the "Search for Extraterrestrial Intelligence," originally financed by the NASA. Unfortunately no trace of extraterrestrial intelligence has been found thus far, but many other projects are being conducted with BOINC software, aimed at solving challenging problems like predicting how proteins "fold" or climate changes occur.

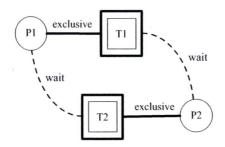

FIGURE 8.3: Example of deadlock in exclusive mode.

a shared bowl of spaghetti in the center, alternating between thinking and eating phases. There are five forks on the table, each put between a pair of philosophers. And now as it is difficult to eat spaghetti with one fork each philosopher must use two forks. We do not know whether this need to eat spaghetti with two forks is a peculiarity of philosophers or if it was an invention for the problem, but in any case each philosopher must grasp the two forks on his right and his left in order to start an eating phase, after which he releases the forks thus allowing others to eat. So far so good if each philosopher waits until both forks are available before starting. But since philosophers are overeager they may grasp any free fork waiting for the second to become available. So the five may end up with one fork each, and none of them will ever eat.[6]

This situation, known as *deadlock*, may occur much more frequently than expected and can be very complex if, as for the philosophers, a block involves a group of entities in a cycle leading to a global paralysis. Among many examples, consider the problem of regulating the access of a process (the entity) to a data structure such as an indexing table (the resource). To avoid inconsistencies in the contents of the table, exclusive access is commonly required during writing operations. In Figure 8.3 two processes $P1$ and $P2$ have gained exclusive access to two tables $T1$ and $T2$ respectively, so access to the tables is not allowed until the operations of $P1$ and $P2$ have been completed. A deadlock occurs if during such operations $P1$ needs some data contained in $T2$, and $P2$ needs some data contained in $T1$. Often a deadlock can be predicted and avoided, in other situations it may be prevented only if the whole sequence of resource requests is known in advance. In some cases, however, this plague cannot be overcome and the computation must be organized differently.

Another difficulty is due to the fact that, while a computation centralized in a single entity can exploit its full knowledge of everything that has

---

[6]The dining philosophers problem was introduced long ago by Sir Charles Antony Richard (Tony) Hoare, a famous British scientist who is rightly considered one of the fathers of computer science.

happened, in a parallel and particularly in a distributed setting, such as the Internet, each entity may have only a partial picture of the activities carried on by the others at the same instant. An entity may not know the total number of participating entities, nor where or what they are. And even if it knows all the features of the network it is hard to foresee how a distributed process will develop because its execution depends on the communication delays that in turn depend on unpredictable network traffic. In general the order in which the different messages arrive at each entity varies unpredictably and affects the way the computation continues. So, even if the same algorithm is used several times on precisely the same input, its execution may vary each time, always producing a correct result but after a different duration. As a consequence the execution time is just an *ideal* measure of the algorithm, because it is given with respect to a hypothetical execution where each operation, communication included, takes a single step. This is why other parameters become important, such as the number or the size of the messages sent during the execution, or simply the popularity of a protocol on the Internet. We will now try to clarify all the above concepts by entering into the world of distributed computation in a unorthodox way.

8.3   A distributed world

Eight secret agents A, B, C, D, E, F, G, H stay in contact by phone using the network of Figure 8.4 which indicates twelve existing direct lines. Each agent knows the telephone numbers of his neighbors with whom he can speak directly, so for example B may call C but at least two calls are needed to send a message from B to G. One day central command issues a communication that only seven of the twelve lines available can be securely maintained, leaving to the agents the task of selecting lines in such a way that all agents will still be able to communicate with each other. In fact as the graph of the network has eight nodes, seven properly chosen lines are sufficient to maintain the connectivity of the group forming a *connected graph* as is shown for example on the right side of Figure 8.4. Note that each communication between two agents may now pass through several nodes thus requiring more calls than before. The order is sent from the central command to one agent elected as the *leader* for this operation, and the leader must arrange for the information to reach the other agents via telephone. To do this the agents must execute an algorithm that in this case is called a *communication protocol* consisting of a set of calls over the original lines, until a set of safe lines has been established.

While executing the protocol, the agents assume one of three possible conditions of *LEADER*, *IDLE*, or *DONE*, and react to the order at different times depending on the lines used, traffic on the network, and personal reaction time. At the beginning all the agents are *IDLE* except for the chosen *LEADER*.

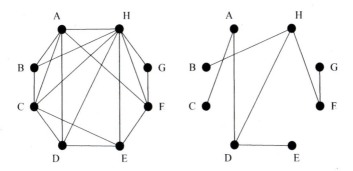

FIGURE 8.4: A communication network with eight nodes and seventeen lines, and a possible set of seven safe lines connecting all nodes to one another.

The operation starts when the *LEADER* calls his *neighbors* (i.e., the agents directly connected with him) to communicate the order. Then the neighbors wake up and start selecting the safe lines. When an *IDLE* agent x receives the order by phone, say from agent y, he accepts the connection and agrees with y that their connecting line (x, y) has to be taken as safe. Then x refuses any further incoming calls from a neighbor that communicates the order to him, and calls all its neighbors except y to communicate the order to them (if one of the neighbors, say z, has already received the order from another line the call will be refused by z). At this point x becomes *DONE*, a status indicating that he has completed his task. The set of safe lines is built when all agents are *DONE*.

Formally the protocol can be specified as shown in Figure 8.5. The protocol is written for an arbitrary number of agents n and refers to a generic agent x because all the agents must follow the same procedure. Some delicate points remain, which we will come to in time, but first examine the given code to make sure that it implements the actions indicated above. Recall that the protocol starts with all agents *IDLE* except for the *LEADER*.

Although the actions of each agent are clearly defined in the protocol, the effect of the overall process may be difficult to understand, as is the case for most distributed processes. In particular we claim that, when all the agents have reached the status *DONE*, a final set of exactly $n - 1$ safe lines has been built. Before trying to prove this, consider a simulation of the agent behavior for the set of connections of Figure 8.4 to see how the protocol works. A number of interesting facts emerge. In particular the final set of safe connections depends on the timing of the phone calls and on the order in which the calls are made, so that different sets may come out for the same graph, starting with the same leader. Note that the operation: *call all the neighbors disregarding any call received*, although apparently parallel, requires that x calls its neighbors one after the other, since a phone call cannot start

algorithm SELECTION

The generic agent x may be LEADER, IDLE, or DONE

if *x* is *LEADER*:

    call all the neighbors disregarding any call received;

    for any neighbor *z* accepting the call:

        add the line $\{x, z\}$ to the set of safe lines;

    become *DONE*;

if *x* is *IDLE*:

    upon reception of a call from another agent *y*

      add the line $\{x, y\}$ to the set of safe lines;

      call all neighbors except for *y* disregarding any call received;

        for any neighbor *z* accepting the call:

            add the line $\{x, z\}$ to the set of safe lines;

    become *DONE*;

if *x* is *DONE*: disregard any call received.

FIGURE 8.5: Protocol for safe lines selection executed by each agent *x*.

before a previous one has ended. Take *H* as *LEADER* and assume that each agent calls the others in circular alphabetic order. That is agent *A* follows the order *B, C, D, F, H*; agent *B* follows the order *C, H, A*; agent *C* follows the order *D, E, H, A, B*; etc. Calls arriving in the same instant are answered in an arbitrary order. Assume that the lines have connection delays of one (e.g., one second) or two, and consider the following two limit cases.

Case 1. The lines of the octagonal perimeter of the graph have delay one and the other lines have delay two. So *H* wakes up *A* at time 1, *A* wakes *B* at time 2, and a chain of calls goes on along the octagonal perimeter waking *G* at time 7. A set of seven safe lines is thus built (graph on the left of Figure 8.6), and all the other calls from each agent, including the calls from the *LEADER*, are refused by his neighbors since they arrive at later times. Different safe lines would have been obtained if *H* had called his neighbors in a different order, even with the same delays on the lines.

Case 2. The lines between *H* and his neighbors have a connection delay of one, and the delay of all the other lines is two. In this case *H* wakes-up the other agents at times 1 to 7 and all the other calls are refused. A new set of seven safe lines is built (graph on the right of Figure 8.6). Note that if *A* had called his neighbors in the order *D, F, H, B, C*, he would have reached *D* at time 3 and *F* at time 5 waking-up these two agents before their reception of the calls from *H* that would have been refused.

What is probably more interesting is that by changing the timing on the lines and the order of the calls *any* connected set of seven safe lines can be

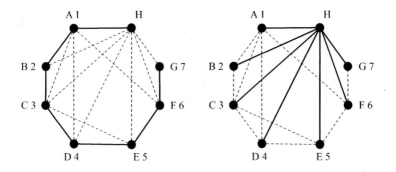

FIGURE 8.6: Safe lines (solid segments) and node wake-up times with *LEADER H*, under two timing assumptions. The lines of the octagonal perimeter have delay 1, or the lines connected with *H* have delay 1, while all the other lines have delay 2.

obtained, as for example the one of Figure 8.4 that the reader may personally investigate.

Although the given protocol is very simple, a formal proof of its correctness can only be given in the realm of the theory of distributed algorithms and is not at all trivial. For those interested in getting deeper into the subject we refer to the bibliographical notes at the end of this chapter, and note here only a few considerations to show how treacherous these algorithms can be. A simple question is whether one can be sure that all agents are reached by the order of the central command. This can be proved inductively. In fact if an agent x other than the *LEADER* were never reached by a call thus remaining *IDLE*, his neighbors would also always be *IDLE*, and the argument can be pushed back to the *LEADER* who is not *IDLE* by definition. A similar argument shows that all the chosen safe lines form a connected graph, but it is more difficult to prove that there are exactly $n - 1$ such lines. The reader may verify informally that the chosen lines induce a tree rooted at the *LEADER*, and recall from Chapter 3 that any tree of n nodes has $n - 1$ arcs, a necessary and sufficient number to keep any graph connected. Aside from the root, the tree is composed of *internal* and *leaf* agents, each of which knows the lines to its parent and the lines to its children, if any. But another question is more delicate to prove.

How do the agents become aware that all their colleagues have become *DONE* and that the set of safe lines is then ready to be used? To achieve this, further actions must be added to the protocol. The information about termination is collected from the leaves to the root on the safe lines of the induced tree, with a so called *convergecast* operation. Then the leader, now the root, becomes aware that all the agents have completed their task and *broadcasts* this information to everybody in a final phase, again using the arcs of the tree. The complete protocol follows as an immediate extension of the previous one but is actually a bit technical and is not reported here. At the end each agent has the knowledge that the selection of safe lines has been completed and can use them with confidence.

The story is much more important than it may appear at a first glance because the protocol is designed for a wealth of problems on arbitrary networks whose entities may not even know the number of their peers, or have a map of the existing connections. If an entity knew the whole network in advance it could select a set of safe lines and communicate this set to the others by itself. In big networks, however, the entities have only a fraction of this information, as in the Internet, which is too big to be known by all nodes, and in which different computers and connections may come up and go down at any time. So the entities present at any moment can only work together according to a distributed protocol based on pieces of local knowledge, like the set of safe lines established by an agent x with his neighbors.

So let us now apply the experience gained with the agents to finding a community in a graph, a problem that was introduced in Section 7.7 of the last chapter. Take again the graph of Figure 8.4 and suppose that some of the agents are interested in a specific topic τ (e.g, movies). One of them, now the LEADER, wants to detect the subgraph of colleagues interested in τ together with all its arcs. This community is easy to detect with a centralized algorithm if the nodes are interrogated about their interests one by one, but they may be so numerous as to render the algorithm impractical, and other interesting features will come out from a distributed approach.

Perhaps surprisingly, the protocol for safe lines selection applies, with a few minor changes. Only the nodes interested in τ now call the neighbors to enquire about preferred topics. They do not disregard incoming calls as was required in the previous protocol for determining a tree with $n - 1$ connections, and retain all the arcs connecting nodes with interest τ as part of the community. The software suite of each node contains the protocol specified in Figure 8.7, that gets activated upon receipt of a enquiry about its interest in the given topic.[7] Note the absolute similarity of the two protocols in Figures 8.5 and 8.7. Assuming A as the leader, two possible communities interested in τ are shown in Figure 8.8.

The protocol terminates when all the nodes interested in topic τ have become DONE. At this point each node in the community has a complete local knowledge of it, that is, the node knows all its neighbors in the community and the arcs connecting to them. This condition can be disclosed to the LEADER during a termination phase, as it was indicated for the protocol of safe lines detection. However, further information on the community can be gathered now. For example the subgraph in Figure 8.8(a) is an α-community (see definition (7.8) in the previous chapter). The subgraph in Figure 8.8(b),

---

[7]Some more details are needed in the protocol to handle the interleaving of outgoing and incoming calls.

algorithm COMMUNITY *The generic node* x may be LEADER, IDLE, or DONE

if x is *LEADER*:
    call all the neighbors;
        for any neighbor z answering **yes**:
            add node z and arc $\{x, z\}$ to the community;
        become *DONE*;
if x is *IDLE* and is interested in τ:
    upon receipt of the first call from another agent y
        answer yes;
        add node y and arc $\{y, z\}$ to the community;
        call all the neighbors except those from which a call was
            received;
        for any neighbor z answering **yes**:
            add node z and arc $\{x, z\}$ to the community;
        for any call received (*during the node's own calls*) from w:
        answer yes;
        add node w and arc $\{x, w\}$ to the community;
        become *DONE*;
if x is *IDLE* and is not interested in τ:
    whenever receiving a call for τ **answer no**;

FIGURE 8.7: Protocol for the detection of a community with interest in a topic τ, executed by each node x. "Calls" implicitly ask for interest in τ. Note that only the leader may disregard any call received.

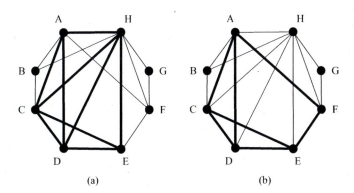

(a)   (b)

FIGURE 8.8: Two possible communities with an interest in τ (see engrossed arcs). (a) A,C,D,E,H in an α-community. (b) A,C,D,E,F is not an α-community due to node F.

instead, is not an α-community because two of the four arcs incident to node F point outside the community. This information, and other properties such as the identity of the participating nodes, can be gathered in a final phase of the protocol.

The peculiarities of distributed systems are quite surprising. Without global knowledge some apparently easy questions may have no answer, while some complex operations can be performed on the whole network using only the local knowledge of the nodes. In the case of the secret agents, when the safe line selection protocol is terminated a global connectivity is established on the whole network, and each agent can use his local safe lines leaving to the agents connected to him the responsibility of using in turn only safe lines with their neighbors. On the other hand it may be difficult or even impossible to solve elementary problems as we now show on a much simpler case.

8.4   Some logically hard problems

The new situation comes from a logic puzzle known as *the two generals problem* born in the framework of the communication between two entities, whose solution is immediate if nothing goes wrong but becomes impossible in the presence of unpredictable transmission faults. Although it is obvious that poorly functioning lines may prevent a sound exchange of information, what is surprising is that the communication may not reach a final state just because transmission faults *are possible*, even if they do not actually occur. Let us see why. Two generals $G1$ and $G2$, each one heading an army, camp at the foot of a hill where a third general E, the enemy, is lodged in a fortress with his troops. A coordinated attack of the two generals $G1, G2$ will defeat E and the fortress will be conquered, while a solitary attack will result in a disastrous failure because E enjoys a better strategic position. $G1$ and $G2$ communicate by sending messengers to agree on a time to attack. So the leader general, for example $G1$, sends a message to $G2$ saying: "lets do it tomorrow at nine o'clock," $G2$ receives the message, and both attack concurrently. No further communication is needed if there is the certainty that the messages arrive at their destination. But how do the generals behave if they fear that a messenger can be ambushed by the enemy? $G1$ can send the same message to B, but before ordering the attack must make certain that the messenger has completed the mission, hence he waits for an acknowledgement from $G2$ e.g., in the form: "I received the message and agree to attack at nine o'clock." Now in turn $G2$ wants to make certain that $G1$ has received his answer before attacking, so $G1$ must confirm once received the message from $G2$, and so on. Since before attacking each general wants to receive from the other an acknowledgment of the last message sent, the exchange will go on for ever

algorithm WITHDRAWAL

> *At ATM terminal T, user U requests to withdraw an amount \$*
> *from bank B.*
>
> T sends a request $M_1 = R(U, \$)$ to B;
>
> *(upon reception of M_1 and if \$ is available)*
> B sends an authorization $M_2 = A(U, \$)$ to T;
>
> *(upon reception of M_2)*
> T pays \$ to U and sends a confirmation $M_3 = C(U, \$)$ to B;
>
> *(upon reception of M_3)*
> B withdraws \$ from the account of U.

FIGURE 8.9: A protocol for bank withdrawal.

(and in any case well after "nine o'clock" of the next day). The attack is then impossible.

This scheme reflects a situation arising in distributed systems, where a problem may be easily solved if everything works well, but becomes impossible to handle in the presence of even intermittent faults. A classical application is in a banking system: a simplified version of a protocol for withdrawal at ATM terminals is shown in Figure 8.9.[8]

This protocol works correctly only in the absence of transmission faults. If M_1 or M_2 fails to arrive the protocol is interrupted with the (minor) consequence that U cannot withdraw the sum required. A failure in the transmission of M_3 is considered more serious because a dishonest user could cash a sum without being charged and the bank B would suffer a loss. To prevent such a case, B could modify its portion of the protocol by making a precautionary withdrawal from the account of U before sending the authorization M_2, and then confirming such a withdrawal upon reception of M_3. In this case, if M_2 fails, the users account is charged without making the corresponding sum available at the terminal, and the user would suffer a loss. It can be easily seen that, as for the two generals, an exchange of messages could go on forever without reaching a state of consistency between U and B, for which a *rendezvous* on different grounds is necessary.

As one may expect the two generals problem has been extended to more than two entities, thus moving from the theory of communications to the world of distributed systems working in the presence of faults.[9]

A fundamental demand is that all entities attain common knowledge on the

---

[8]Actual ATM terminals work differently. See the bibliographic notes at the end of the chapter.

[9]The extended version, called the Byzantine generals problem, is far too complicated to be presented here. The term Byzantine refers to protocols where different faults may be generated by a malicious "adversary" with the purpose of subverting the process. The rationale behind this approach is that a protocol resistant to Byzantine faults would resist any faults.

system in a limited number of steps, because the success of certain cooperative actions may depend on it. Clearly this possibility depends on the parameters of the problem and in many cases is not attainable. More generally there is a subtle interdependence between knowledge, communication, and action. An additional complication comes from the lack of reflexivity of certain relations, as typically occurs in a directed graph where a node A may point to node B but the converse is not necessarily true. This raises new challenges in information propagation because B cannot send a message to A if a path of directed arcs from B to A does not exist, so that achieving common knowledge on the whole system may become hard for all the entities. Lack of reflexivity occurs in a wealth of different situations, and different solutions have been proposed from case to case, some of them being really surprising. Among others, a brainteaser pertaining to a non-reflexive world known well before the advent of computer networks is frequently used to understand how common knowledge can be attained in a distributed system by other means than mere message exchange. We call it the *problem of the jealous Amazons*. Its tricky solution is based on the known fact that Amazons are very smart and all of them assume that the others are as smart as they are (the associated implicit assumption is that the entities in a network are as smart as the Amazons).[10]

In the country of Amazons, just as elsewhere, when somebody has an unfaithful partner, everybody knows about it except for him/her-self, a clear lack of reflexivity. According to a version of the myth no men were permitted to have sexual encounters with Amazons or reside in their country, and female children were adopted from other tribes and brought up as future Amazons to prevent their community from dying out. Nevertheless, when visiting a neighboring tribe for adopting children, the Amazons were exposed to heterosexual temptations that occasionally resulted in their habitual female partners being cheated. Of course every Amazon knew immediately of their cheated colleagues except for the cheated ones.

But enough is enough. One day, after the visit of her subjects to a tribe of handsome men, the Queen of the Amazons proclaimed a firm resolution in Main Street:

"In this country there are unfaithful Amazons. For the sake of social order all further visits abroad are suspended until morality is fully restored in the kingdom. It is not permitted to communicate on this issue in any way, however, as soon as one of you is certain that her partner has had an affair outside of the couple, you shall kill her on that precise day."

The Amazons went back to their activities. Knowing the Queens severity, none of them ventured to speak or even to mention the problem, although all other rumors were immediately spread as usual. It turns out that there were

---

[10]The problem is generally formulated on "cheating wives"(see the bibliographical notes), but is also found under different names, with different formulations. We adopt a "jealous Amazons" version which refers to more open relations.

thirteen unfaithful Amazons. Twelve quiet days went by, but in the morning of the thirteenth day, thirteen arrows pierced the hearts of the culprits. How was this possible? The question is interesting because *apparently* the Queens speech did not add anything to what everybody already knew, that is that some Amazons had been cheating. We will return to this point later, noting that a clear novelty lies in the triggering order that specifies that action has to be taken in a given interval of time ("that precise day").

First note that the result, valid here for thirteen Amazons, is valid for any number k of Amazons and the rule becomes: "*if there are k unfaithful Amazons, in the $k - th$ day all of them will be killed.*" It is convenient to start examining the question from small values of k and then generalize the reasoning to arbitrary values. If $k = 1$ there is only one unfaithful Amazon and everybody knows it except for her partner. This poor lady, not having heard before about infidelity, immediately understands and kills her partner in the same day. Here is the point where the apparent neutrality of the Queens speech fails, because her assertion on the existence of cheating Amazons is new information for the cheated one. If $k = 2$, the two cheated Amazons know about one unfaithful partner and wait with anxiety until midnight of the first day to know if the culprit that they know has been killed (recall that all rumors spread immediately). Such a piece of news would have confirmed that they had complete knowledge of the kingdom, that is, there was only one unfaithful Amazon. However no news arrives in the morning, so they understand that the unfaithful ladies are indeed two including their own partners, and in the second day both culprits are killed. By induction, k cheated Amazons know about $k - 1$ unfaithful partners. They wait up to midnight of the $(k - 1) - th$ day and start looking for news. As none of the cheating Amazons they know about is reported as being killed, they understand the situation and, on that $k - th$ day, shoot an arrow into their partners' hearts.

Note that for any Amazon the uncertainty is only between two values k and $k - 1$ and the problem can be solved because the Amazons are capable of quick logic reasoning (they must be able to react within twenty-four hours). Furthermore, although Amazons may cheat sexually they are honorable warriors. Having full knowledge of what happens in the kingdom except possibly of herself, each one of the cheating Amazons knows that eventually she will be executed but does not try to escape her just punishment (e.g., by fleeing to a neighboring tribe). The argument for establishing the day of ones own execution is left as an exercise. The problem is subtler than it may appear because a couple of Amazons may have been mutually unfaithful and, as honorable subjects, they face each other and cast the arrows at the same time.

The jealous Amazons paradigm is relevant, for example, in designing computer network protocols where a knowledge of the system leading to certain actions may be acquired with message exchange, but also through the examination of particular events that may occur. To make it clear, let us look in some more detail than we have considered so far at the diffusion of messages in a network.

8.5   A closer look at routing

If a network is represented as a graph, a connected sub-graph including all the nodes and a minimum number of arcs is called a *spanning tree*. As for any tree, if the graph has n nodes a spanning tree has $n - 1$ arcs. In the secret agents problem exactly seven lines were required for connecting the eight agents. All the different sets of safe lines indicated before are in fact spanning trees for the agents graph, showing that there may be many spanning trees for the same graph. If a *weight* is associated to each arc, e.g., the value of the transmission delay or the cost of the line rental, it may be important to determine a *minimal* spanning tree in which the sum of the arc weights is minimal. For example the two spanning trees of figure 8.6 are minimal if the communication delays are taken as weights. Note that several minimal spanning trees may exist for the same weighted graph, although this is not the case for Figure 8.6.

Determining a spanning tree in a network is important for maintaining the connection among a group of entities using the minimum number of lines. The messages travel from one entity to a neighbor with one *hop* on the corresponding arc of the tree, and the overall communication takes place according to different strategies. In the *broadcast mode* a message is sent in separate copies from one entity to all its neighbors in the tree. These in turn broadcast the message to their neighbors (except the one from which the message was received) and the process goes on until all the entities have been reached. This mode is employed when all the entities have to acquire some common knowledge as in the termination phase of the agents protocol; or when all the entities must participate in a common action; or when it is not known which entities need the information or where they are. Note that a broadcast produces $n - 1$ copies of the original message, that is as many as there are arcs in the tree, so the mode must be used sparingly so as not to cause congestion in network traffic.

A different communication mode called *routing* is adopted when the final destination of a message is known. For example in the Internet the address of the final receiver travels together with the message (i.e., *the packet*), which proceeds to its destination as a single copy, along a path of minimal weight that accounts for the total transmission time and for the cost and reliability of the lines involved. The mechanism used to determine this path is very complex and is performed before the routing starts. We can only give an intuition of how it is accomplished here. Each node stores a set of *routing tables* that specify, for any possible final destination (*dest*), the selected neighbor (*next*) to which a traveling message must be sent to follow a path of minimal weight.

Consider again the agents graph where the lines of the octagonal perimeter have weight 1 and all the other lines have weight 2. The graph is repeated in Figure 8.10 with the routing tables for the nodes C and F (now all the lines are

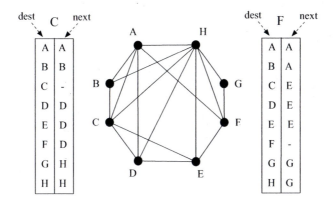

FIGURE 8.10: Routing tables for the agents C and F in a graph where the lines of the octagonal perimeter have weight 1 and all the other lines have weight 2. For each final destination (dest) of a message, the table shows the line to be taken (next).

available for use). Note that these tables store only local information, leaving to the selected neighbor the task of deciding the next step. For example, the table stored in C indicates that a message arriving in C and directed to F has to be routed first to D, leaving to node D the choice of the next arc in the path. The routing tables are not uniquely determined if more than one path of minimal weight connects two nodes. For example the path $C - D - E$ has weight 2 as well as the direct connection $C - E$, hence in the table of C we could enter E instead of D, for destination E.[11]

Building the routing tables is a major task in network operation, as well as keeping them up to date if substantial changes occur in the network. The protocol devoted to it requires that each node transfers to its neighbors its knowledge of the shape of the network, and this proceeds in broadcast mode. In a highly dynamic network like the Internet the protocol is executed periodically because nodes and arcs are inserted or removed continuously and transmission delays change with the network traffic. Still, the tables cannot be changed at just any moment and the nodes have to be able to cope with possible blockages or delays on the lines. This is where the alliance of knowledge, communication, and action shows its role. Without getting into complicated details, refer to Figure 8.10 again and assume that node C finds that the connection $C - D$ is temporarily blocked, e.g., because a long sequence of

---

[11]Due to the enormous size of the Internet it is not practical to have tables that store all the network destinations. As we know from Chapter 7, the Internet is subdivided into sub-networks. Routing of a message is accomplished by jumping from one sub-network to another, which is in turn traversed with a similar strategy. A routing table then refers to a specific level of routing.

messages in the path $A - D - E$ are queued in D. After time-out, C changes the routing to F sending the messages through H instead of D, although the new path may have higher weight (4 against 3).[12] Note that such a decision cannot prevent a message from reaching its final destination because each vertex (H in this case) can send messages everywhere. Now, when F observes that the messages routed by C are not arriving from E as usual, it may infer that there is a blockage somewhere in the standard path. Just like a jealous Amazon, F obtains a piece of global knowledge by examining the behavior of a peer. The result is that F may decide to send in turn messages to C through H (or maybe through A etc.) instead of through E although the connection $F - E$ that it can directly monitor is working properly.

The field of routing is huge and difficult. In this section we have provided just an introduction: the interested reader is advised to refer to the following biographical note.

Bibliographic notes

To investigate the world of parallel and distributed computation it is advisable to start from books offering coverage of the whole area. The classical text of F. Thomson Leighton, *Introduction to Parallel Algorithms and Architectures: Array, Trees, Hypercubes*, Morgan Kaufmann Publishing, 1991, is still the most general in scope, and is easily understandable by an audience at college level. A must in distributed computing is *Distributed Algorithms* by Nancy Lynch, Morgan Kauffman Publishing, 1996. However we strongly recommend a recent book by Nicola Santoro, *Design and Analysis of Distributed Algorithms*, Wiley Interscience, 2007, written in a rigorous but witty style.

Following the development of the chapter we start by remarking that Kevin Kelly is an influential journalist who has written on many subjects. In the world of computer science he is particularly known for the attention devoted to the activity of hackers since the early days of the Internet, and for having founded and edited the brilliant magazine *Wired* where technology and culture are combined. Readers interested in the impact of the Internet on our society should take a look at this magazine, at least occasionally.

Information on grid and cloud computing is easily found by browsing the Web. A good reference on cloud computing in general is: Avetysian, A.I., *et al.* 2010. Open Cirrus: a Global Cloud Computing Testbed. *Computer*, Vol. 34-4, pp. 35-43. IEEE Computer Society. Furthermore, the whole issue of *Computer*, Vol. 44-3 (March 2011), is devoted to cloud computing. Information on projects centered on open distributed computing, including BOINC and SETI, can also be found in the Web.

The two generals problem was formally described for the first time in 1975 by E. A. Akkoyunlu, K. Ekanadham, and R. V. Huber, "Some Constraints

---

[12]If this new choice is stored in an alternative table, the routing information kept at C becomes quite high. If C has no knowledge of the cost of the possible alternatives it may select the new destination at random.

and Tradeoffs in the Design of Network Communications," and later applied to database consistency. To discover how bank terminals work one may refer to Chapter 10 of Ross Anderson's "Security Engineering," published by Wiley and generously made available on the Web.

An official formulation of the problem of cheating wives is found, e.g., in: G. Gamos and A. Stern, *Forty unfaithful wives. Puzzle Math*, The Viking Press, NY 1958, pp. 20-23. For the connection between this problem and distributed systems see in particular: Y. Moses, D. Dolev, and J. Y. Halpern, Cheating husbands and other stories: A case study of knowledge, action, and communication, *Distributed Computing* 1 (3) 1986, Springer, Berlin.

Routing must be studied in textbooks on computer networks like for example: Tanenbaum, A.S. 2003. *Computer Networks*. IV Edition. Pearson Education. To go deeper into routing algorithms, a good reference is the Santoro book mentioned above.

Chapter 9

Browsers and search engines

How data can be extracted from the Web in an astonishingly short time, without any certainty that what we get is what we actually wanted.

When the "network" first appeared at the end of the 1960s, only a small clique of experts could use it. ARPANET had been designed by keen scholars as a military and research tool, without the intention of it ever becoming a vehicle (perhaps *the* vehicle) for commercial business. And even after the transition of the 1980s when ARPANET and other existing networks started operating together under the TCP/IP protocol giving rise to the Internet, the network was still not suitable for the general public. The killer idea was the creation of the Web, with the crucial feature of assigning a Uniform Resource Locator (URL) to each Web page, for example: http://en.wikipedia.org/wiki/URL - the Wikipedia page about URLs. But even then, wandering in the ever growing Web was hard for non specialists, a little like navigating an ocean without GPS or even a compass.[1] A great improvement in Internet usage came with the development of browsers that permit any Web page to be viewed once its URL is known. The search can start from a given or previously known site, and proceed following clickable links until a desired or at least interesting page is reached. It goes without saying that most of the relevant information on the Internet remains unreachable if one can only use a browser in this way, but the big leap towards the consumer Internet revolution had been taken.

The first popular browser was Netscape Navigator, developed from the earlier prototype Mosaic produced by the University of Illinois and released freely to non-commercial users at the end of 1994. Compared to earlier software, the major innovation of this new browser was that it allowed on-the-fly display of web pages, a feature of great importance at the time because connection speed was generally very limited. A user could start reading a page in seconds as long as a portion of data had arrived, without waiting for the full loading of the page. Graphics could be made to load last. The Netscape browser was compatible with most operating systems and remained the tech-

---

[1]This analogy may explain the proliferation of deceptive marine terms on Web searching like navigation, surfing, crawling (if we consider it a method of swimming): all operations made by a stationary person on a stationary computer via stationary software tools.

nical leader during the decade. Many other browsers then appeared, among which Microsoft Explorer and Safari became very popular. Netscape does not exist any more, but some of its code is still used in its open source successor Firefox, one of the best browsers in operation today.

9.1   Caching Web pages

One can easily imagine that designing a web browser requires great skill in computer and communication technology. Many of the problems encountered are of an algorithmic nature. Let us mention one of them, crucial for an efficient functioning of all browsers, that falls into the NP-hard class of intractable problems discussed in Chapter 4.

Although most of the communication bandwidth of the Internet seems to be taken up by music and video sharing, and this tendency is ever-increasing since the introduction of high definition videos and 3D films, the downloading of Web pages is responsible for a sizable proportion of bandwidth consumption. Limiting page transport is therefore beneficial for the efficient usage of the Internet overall, and also allows browsers to respond faster. To this end frequently requested Web pages are copied and *cached* in various locations, from single computers to whole dedicated subnetworks, and then retrieved from the closest location holding a copy of the page required.

Part of a caching infrastructure is shown in Figure 9.1, limited to a large autonomous system and its connection to the Internet. To reduce the traffic through the router and inside the local network, and to avoid infiltration of possibly dangerous files, *proxy* caches keep copies of the pages most frequently required by the browsers connected to them. If two or more browsers require the same page from outside the AS, this is loaded into the proxy and then extracted from there. If an AS page is frequently required from outside it is also stored in the proxy so the local network is minimally affected. Even the exchange of pages between computers of the same AS may be speeded up with proxy caching.

Caching, however, is not limited to the local networks of the ASs. Different cache infrastructures have been deployed in the Internet until, from the year 2000, very large *content distribution networks* (CDN) have appeared, capable of serving as storage for whole geographical regions. This has given rise to a wealth of optimization problems aimed at deciding which pages have to be replicated, and where they should be stored. Let us consider an extreme simplification of one such problem in order to show that, even in this minimal form, exponential time is required for attaining an optimal solution.

Assume that only two computers exchange Web pages in the oversimplified network shown in Figure 9.2, with two routers and one CDN cache. Pages have different sizes and the CDN cache can allocate them up to a total size 18,

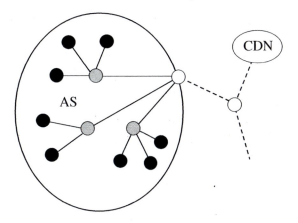

FIGURE 9.1: An autonomous system connected to a CDN through the Internet. Computer terminals (containing a browser) are shown in black. Proxies are grey. Internet routers are white.

insufficient to store all pages. If a computer requires one page from the other, and this page is stored in the cache, then only one server must be traversed, otherwise three hops are needed. Assuming that the hop time is proportional to the page size, and that all pages are requested with the same frequency, the cache should be loaded with pages of a total maximal size. If the page sizes are 3, 9, 10, 4, 7, 12 as shown in the figure, two optimal solutions are caching 3, 4, 10, or 7, 10, for a total cache utilization 17 (note that the full capacity 18 cannot be reached). But how can these solutions be found?

The problem that we are facing is well known in complexity theory where it goes under the name of *subset sum*. Given a set A of n positive integers, and another integer k, determine a subset of $B \subseteq A$ such that:

$$S = \sum_{b \in B} \leq k \tag{9.1}$$

and no other subset of A has a sum S' with $S < S' \leq k$. In our example we had $A = \{3, 9, 10, 4, 7, 12\}$ and $k = 18$, and $B_1 = \{3, 4, 10\}$ and $B_2 = \{7, 10\}$ were two possible solutions.

Unfortunately subset sum is exponentially difficult, and in fact is an NP-hard problem as introduced in Chapter 4. It has a perfectly simple formulation but is extremely hard to solve and no algorithm polynomial in n is known for it. As for all the problems of this class no efficient strategy is known, so we must essentially rely on a complete enumeration of all possibilities, in this case of all the subsets of A, to find the one that meets the conditions required. And since a set of n elements has 2^n subsets, we must examine $O(2^n)$ cases even though many of them can be discarded beforehand for trivial reasons (e.g., they contain an element greater that k, etc.). This point deserves some more attention.

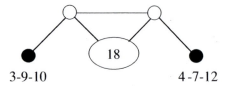

3-9-10   4-7-12

FIGURE 9.2: Allocation of Web pages into a CDN: a limit case based on subset sum. Pages of size 3, 4, and 10 go into the memory bank of size 18.

The example developed in Section 4.5 was about finding a special community called k-clique in a graph of n nodes. Without a strategy all the $\binom{n}{k}$ sub-graphs of k nodes must be examined, a function upper bounded by n^k, then of order n^n if k grows linearly with n. For the subset sum, instead, the number of probes is "only" 2^n (for example for a set of thirty elements the number of probes is a little more than one billion). Notwithstanding this great difference between the two mathematical functions, it can be proved that the two problems (like all the NP-hard problems) are strictly connected to one another, in the sense that discovering a polynomial algorithm for solving one of them would automatically be the key to a polynomial solution of the other. Theoretically we are in a really challenging world. But how can we do page caching in practice?

Clearly the starting case of Figure 9.2 means nothing in the real world, but it still carries important conclusions if we extend it to practical cases. First we can have an arbitrary number of computers exchanging pages through an arbitrary network, with an arbitrary number of routers. Since the problem is exponentially difficult in the starting case, it remains at least as difficult in the extension that admits the previous one as a special instance. Similarly an arbitrary number of caches of different levels can be introduced in the network, and the problem will be at least as difficult as before. However, an even more important extension must be considered.

The real benefit of caching pages is not only related to their size, but to the distance between the original location and the source of the request. Copying a page into a CDN or into any other cache infrastructure may reduce the number of its hops in the networks substantially: so a "value" $v(p)$ is assigned to each page p, reflecting the overall saving if the page is cashed somewhere. $v(p)$ can be expressed by a complicated function that takes into account the number of copies of p, their locations in the network, and the frequency and sources of the requests of p if known or foreseeable. Clearly this approach induces further complications, so the problem remains exponential. In fact it can be formulated in terms of another well known NP-hard problem known as *knapsack*, a sort of extension of subset-sum where the elements of the set A also have a value, and the aim is maximizing the sum of the values in the chosen subset B without exceeding the overall size k. Since something must be done anyway, several heuristics have been proposed for the Web cache problem as indicated in the bibliographical notes.

9.2 From browsers to search engines

Browsers constituted a fundamental advance in the usability of the Internet. However, the great innovation that allowed many people without computer skills to enter the Internet world was the *search engine*. The major ones today are Google, Yahoo!, Microsoft's Bing, and the spectacularly fast growing Baidu in the Chinese language. We are all used to them and tend to forget the difficult user experience of the first search engines in the 1990s, when, after a stock-market enthusiasm driven by speculation, many companies were hit by the bursting of the "dot.com bubble."

The pioneering search engines often created more problems than they could solve, by returning exceedingly long lists of information without any apparent rationale behind them. In fact these engines were based on some principles of information retrieval that, although well established and sound, were insufficient for making a satisfactory selection among all the possible answers if not combined with other criteria. Very often the queries had to be given in an expert way, and refined in successive, answer-driven steps, to have any hope of success. Searching was a job for programmers or database experts, while ordinary people were lost in a sea of irrelevant results.

Around 2000, however, Google arrived, and achieved spectacular success in a very short time.[2] Indeed at that time an increasing number of scholars, programmers, and engineers were studying how to provide user-friendly access to the Web and many results of this work were already available. What Google introduced, a crucial innovation, was the algorithm *Page Rank* based on the mathematical concept of a Markov chain, that assigned a "popularity" score to the Web pages according to the number of incoming links in the Web graph, returning the pages with highest score among the ones containing the keywords of the query. Around the same time Jon M. Kleinberg, a professor of Cornell University on leave at IBM research, introduced the *Hyperlink induced topic search* (HITS) algorithm based on completely different concepts but still exploiting the structure of the Web graph. We will explain the two approaches in detail in the next section.

It is customary, however, to tell of the inception of Google as a fairy tale. Two young and very clever students, Sergey Brin and Larry Page, developed the engine as part of their academic work at Stanford University. The name

---

[2]The arrival of Google has been even defined a "black swan," that is, a highly improbable event with tremendous consequences. See: Taleb N.N. 2007. *The Black Swan*. Penguin Books. As we discuss above there was nothing improbable about Google or any other search engines development, as they all used a wealth of public studies and results.

was chosen as a rewriting of *googol*, a word born from the mind of a little boy who was asked by his grandfather to give a name to a number composed by a one followed by a hundred zeros. This is more than the number of stars in the sky, but the name sounds also similar to goggles, the aquatic lenses needed to look into the vast ocean of the Web. The company started operating in 1998 in a garage in Menlo Park, California, making use of open source software.

After the success of Google all other search engines started using properties of the Web graph to improve the quality of the answers through page popularity, so that almost all engines today obtain comparable results leaving the competition to be fought out on other grounds, in particular on the variety of different services offered. Discussing these aspects is outside the scope of this book. We only remark that popularity is not related to the *quality* of a Web page, in the same way a movie can be very popular due to the participation of some famous actors and strong marketing, independently of any real value. Therefore the exercise of our own critical faculties, together with a verification via different sources, remains essential in assessing the result of a search.

9.3   The anatomy of a search engine

The survivors of the previous chapters will now be embarking on an important part of this book: namely, the description of the construction and operation of a search engine. This is a very technical topic and some of the proprietary solutions actually implemented are kept jealously confidential by the search engine owners. What we shall do here is to explain as simply as possible the basic features common to all major engines, and what these features have to do with the algorithmic and mathematical concepts discussed thus far. Or, in other words, how the miracle of obtaining thousands of answers in a second from a search engine is possible, arranged more or less in order of decreasing importance.

To this aim we divide our discussion in five parts, respectively treating the basic data structures and algorithms used; how data are extracted from the Web and stored into the engine memories; how the relative importance of the collected pages is decided; how user queries are answered; and how the use of distributed techniques and parallel processing make the final result possible.

9.3.1   The basic data structures

Search engines collect an enormous amount of data from the Web that must be organized into memory and kept ready to be sent back to the users in answer to their queries.[3] The basic data structures used are extremely large but not particularly sophisticated, consisting mainly of arrays and trees. What is crucial is how the insertion and retrieval of data is performed, because the slightest inefficiency becomes dramatic due to the size of these structures. It is not only the case of using polynomial time algorithms, as is obvious, but also weighing up the relative efficiency of such algorithms.

Data are organized in an *inverted file* system, mainly consisting of three huge tables: a **Document table D**; a **Term table T**; and a **Posting table P**. We shall give an idea on how these tables are used in the over-simplified example of Figure 9.3. Their construction is referred to as *indexing*.

The collected pages get an integer identifier **docID** and are stored in D by increasing docID number. The example refers to a site on the Beatles' original compilations remastered (docID = 5); the site Submarinechannel on animation (docID = 20); and a site with the story of the song Yellow Submarine (docID = 90). Both URL and full page text are stored, the latter not shown in the figure.

The terms present in all the documents are stored in table T in alphanumeric order. Terms are words in all natural languages (and their misspelled forms if any), acronyms, e.mail addresses, etc., in fact all strings of characters delimited by blanks found on the Web.[4] The posting field in T for a term t indicates the position i in table P where a list of references to the documents containing t starts. In the example the term "beatles" pointing to $i = 10$ is found in document 5 where it occurs 8 times; in document 90 where it occurs 6 times; etc. The list terminates with a special symbol $ in position $i = 32$, meaning that $(32\text{-}10)/2 = 11$ documents contain the term beatles, two entries of P per document. The name 'inverted file' given to the structure comes from database technology and has to do with the fact that terms are collected by scanning the documents but are retrieved in the order dictated by T.

As one may well imagine, the overall data structure is more complicated than the one shown here. In particular P usually contains not only the number of occurrences of a term in a document but also the positions of such occurrences that, as we shall see, are important for deciding the relevance of the page. Compact data representations are employed using the Huffman prefix code discussed in Chapter 3 and a wealth of other methods, in particular for coding the large integers appearing in the tables. URLs may be substituted by much shorter hash images as explained in Chapter 5. More importantly, the tables D, T, and P may be represented as arrays, trees, or other structures de-

---

[3]When discussing search engines it is customary to mention the number of Web pages collected by an engine, the size of the internal data structures, etc. We refrain from giving numbers that change (in fact, increase) every month. Keep in mind that, in the year 2010 when this book was written, tens of billions of Web pages might be collected by a single search engine, containing hundreds of millions of distinct terms, all of which must be stored for successive processing.

[4]Table T has probably reached an almost stable size and will increase only when aliens start adding their pages to the Web. In the year 2010 we are talking of many hundreds of millions terms.

D: DOCUMENTS

docID	URL
...	...
...	...
5	thebeatles.com
...	...
...	...
20	submarinechannel.com
...	...
...	...
90	en.wikipedia.org/wiki/Yellow_Submarine
...	...
...	...

T: TERMS

term	posting
...	...
...	...
beatles	10
...	...
...	...
submarine	70
...	...
...	...

P: POSTING LISTS

i	...	10				32	...	70				82	...
p	...	5	8	90	6	...	$	...	20	1	90	4	...	$	...

FIGURE 9.3: Inverted file indexing: a simplified example. The collected documents are stored in a table D by increasing docID number. All the terms present in the documents are stored in a table T, where the posting field for a term indicates the position in a posting table P where a list of references to that term start. For example "beatles" in T points to position 10 of P where we find docID = 5 with 8 occurrences of the word beatles in the document of URL thebeatles.com; docID = 90 with 6 occurrences, etc., up to the list terminator $ for beatles in position 32.

rived from these, where the search of elements can be done with very efficient algorithms as BINSEARCH or the equivalent tree search method discussed in Chapter 4. The actual implementations adopted by the different engines are not always known.

Part of the difficulty is that most of the data are stored on disks whose access is much slower than main memory, so the algorithms must also seek to minimize the number of disk accesses. To this end data compression helps substantially, because the total amount of data on disks is reduced.

9.3.2 Crawling the Web

The next aspect of search engines to consider is how data are extracted from the Web. This is the role of *crawlers*, that is computer programs charged with retrieving as many Web pages as possible.[5] Due to the size of the network, each engine actually uses a very large number of crawlers, that work in parallel and interrogate the Web relentlessly.

First note the essential difference between a browser and a crawler. The former resides in the user's computer and is designed to retrieve Web pages whose URL is known. The latter resides in a search engine computer and is designed to collect "all" pages available, with a limitation imposed by the engine policy that demands a minimum threshold of presumed significance. In fact the effort needed for crawling the Web and managing the search engine data structures requires a trade off between the amount of information that can be made available to the users and the speed of operation of the whole system. Let us consider this aspect first.

The number of Web pages is huge and continuously changing, not only because sites are born and die with incredible frequency, but also because many sites change their contents continuously. While it may be important to visit weather report sites rather frequently, so Internet users may have an updated forecast for their home town, it may not be that interesting to collect high atmosphere wind speeds every hour. Even more importantly, Web sites are often organized as whole subgraphs of pages of different "levels," going down from level to level with the addition of a slash into the page URL to reach more and more specific information (for example, en.wikipedia.org is at the first level, en.wikipedia.org/wiki is one level down, etc.). Pages of very low level are said to constitute the *deep Web* and are not collected by the crawlers, as they can be reached anyway if the engine returns a page of higher level pointing to them. All the non-deep pages form the *indexed Web*.

The basic algorithmic structure of a crawler is indicated in Figure 9.4. The program makes use of two data structures, a *queue* called *QUEUE* and two tables *A, B*. By definition the queue keeps its items one above the other, outputs the top item upon request (*QUEUE* → *x*, where the variable *x* takes

---

[5]Crawlers are also called spiders or robots, although the first term is more commonly used.

algorithm CRAWLER 1

  starting condition: $URL_1,..., URL_s$ are in $QUEUE$;

  while $QUEUE \neq \emptyset$ {

    $QUEUE \rightarrow URL$;

    if ($URL \notin A$) {

      request $TEXT(URL)$;

      $URL \rightarrow A$; $TEXT(URL) \rightarrow B$;

      forany link L **in** $TEXT(URL)$ {

        let L point to URL' ;

        if ($URL' \notin A$) $URL' \rightarrow QUEUE$; } } }

FIGURE 9.4: Basic algorithmic structure of a crawler.

the value of the output element), and accepts new elements at the bottom ($x \rightarrow QUEUE$). The two tables can be implemented at will, provided fast lookup and insertion operations are possible.

As the figure shows, a group of s URLs of potentially important sites is initially loaded into the queue as a seed. The crawler asks for the pages with the URLs in the queue and, if not already present in the tables, retrieves both the URL and text and stores them into A and B. It then scans the page just found for possible links contained in it and, if the URLs pointed to have not been encountered yet, it loads them into the queue. The algorithm is very simple and goes on until there are no URLs to be examined into the queue (the termination command **while** $QUEUE \neq \emptyset$ checks for a void queue). In principle a whole connected subgraph of the Web is visited and the algorithm may continue forever if sites keep on changing. In practice the story is different.

As we have seen crawlers are not designed to retrieve all Web pages. Moreover they have to return frequently to the same page if this gets updated continuously, as for example for weather conditions and news, but might visit only once in a while corporate pages that tend to remain stable. Therefore the queue is substituted with a *priority queue* where each element has an associated priority (in fact, an integer). The structure keeps the element of highest priority at the top and returns it upon request. The construction of such a queue can be found in any textbook of data structures: we only note that the elements must be kept in a way that the next top element can be readily identified after the extraction of the top one, hence the new entries must be allocated in their proper positions to make this possible in short (logarithmic) time. Items with the same priority are kept in arrival order.

A new version of the crawler that exploit URL priorities is given in Figure 9.5. The algorithm is not trivial and is given for courageous readers to study. $PQUEUE$ is the priority queue: the way priority is computed depends on the engine policy. The value "refresh" given to priority indicates that the page

algorithm CRAWLER 2

    starting condition:

    (URL_1,P_1), ..., (URL_s,P_s) are in *PQUEUE* ;

    while $PQUEUE \neq \emptyset$ {

        $PQUEUE \rightarrow (URL,P)$;

        if ($URL \in A$ **and** $P =$ *refresh*) {

            request *TEXT(URL)*;

            TEXT(URL) $\rightarrow B$; [replace old occurrence]

            after delay Δ $(URL,P) \rightarrow PQUEUE$; }

        if ($URL \notin A$) {

            request *TEXT(URL)*;

            $URL \rightarrow A$; *TEXT(URL)* $\rightarrow B$;

            forany link L **in** *TEXT(URL)* {

                let L point to URL' ;

                if ($URL' \notin A$) {

                    compute P' for URL' ;

                    if ($P' \geq$ *threshold*)

                      $(URL',P') \rightarrow PQUEUE$; } } } }

FIGURE 9.5: The algorithmic structure of a crawler with priorities.

must be read again so it is returned to the queue after a given delay Δ. A "threshold" is also specified, below which the new page is not fetched.

To make our algorithm even closer to reality we must add some further considerations. First, Web sitemasters may decide not to allow some of their pages to be fetched by search engines. This is specified in a special file stored in the site that the crawler must interpret in order to decide which pages to take. Furthermore crawlers should try to avoid loading different copies of the same page that are repeated at different URLs (as occurs quite often), or pages that are very similar. Hashing whole pages for comparison is a helpful technique for discarding duplicates. Finally crawlers are designed not to pass over the same pages too many times to avoid overloading a site. All these features must be implemented into the crawler algorithm.

A natural question to ask at this point is how one crawler can visit the entire Web in a reasonable amount of time. The answer is that this task is carried out by a large number of crawlers working in a distributed fashion as explained below.

9.3.3 Page relevance and ranking

The tables of URLs and pages filled in during the crawling phase are the initial data used for building the inverted file structure of Figure 9.3 in what is called the indexing process. Other fundamental information is needed, related to the expected importance that each page will assume for an Internet user among the thousands of pages containing the required keywords. This leads to a *ranking* of the retrieved documents, a feature that differentiates particular search engines from one another. Many of the criteria used for ranking are kept secret and evolve continuously: what we report now is a public knowledge of methods used by all major engines.

The first feature to take into account is the *relevance* $R(p,t)$ of a page p as a function of any particular term t occurring in it. For this the positions of the term in the page are important, with the occurrences of t in the URL, or in the title, or in the first lines of p being assigned a greater importance than the other occurrences. But, perhaps unexpectedly, particular occurrences of t in other pages pointing to p are also very relevant for p. In fact a page p' may point to p through a clickable sequence of words called an *anchor text* for p. For example a Web page p' on songs of the 1960s may point to the page p on remastered Beatles' songs through an anchor text *Beatles' compilations* associated to the URL thebeatles.com (this is easily done by the designer of p' in the HTML description of this page). So even the word $t = $ compilations (and its prefix compilation that, together with the plural form, is certainly present in the term table T) becomes relevant for p, besides being relevant for p'. Both $R(p,t)$ and $R(p',t)$ are then affected.

All search engines treat anchor text terms as very relevant both for the pointing and the pointed page, presuming that they have been chosen by the page designers as particularly descriptive of the situation at hand. If many pages point to p using a same term t in their anchor texts, t becomes very relevant for p and the value of $R(p,t)$ increases substantially. This is the first example of relevance decided by "popularity" instead of relying on the intrinsic value of a term (whatever that means).

Besides considering the positions of a term t on a page p, relevance is also affected by the number of times t occurs, and by the significance that t may have in the given context. For example common linguistic elements such as articles and prepositions trivially occur very often and count for practically nothing as distinguishing elements of a page. More interestingly, the term "song" in the page thebeatles.com must be treated as much less relevant than the term "crawling" if the latter occurred in the page, because the appearance of an unexpected term is likely to indicate an important feature of the document at hand.

These concepts were well known in information retrieval long before search engines existed. A commonly accepted metric, called TFIDF for Term Frequency combined with Inverse Document Frequency, is based on the score:

$$S(p,t) = TF(p,t) \times IDF(t) \qquad (9.1)$$

where the term frequency *TF(p,t)* is the number of occurrences of *t* divided by the total number of terms in *p*, and the inverse term frequency *IDF(t)* is related to the unexpectedness of *t* in *p*. Actually *IDF(t)* has a more flexible definition because it is measured with respect to a collection *C* of pages on topics close to the topic of *p*, that may be chosen with a certain freedom (for example *C* is a collection of pages about songs, and *t* is "crawling"). For a given collection *C*, denoting by C_t the elements of *C* containing the term *t*, and by $|C|$ and $|C_t|$ the number of elements in the two collections, the standard definition is then:

$$IDF(t) = \log_2(|C|/|C_t|). \tag{9.2}$$

Note that the value $|C|/|C_t|$ gives an indication on how "strange" *t* is for the group of pages in *C* (in fact, infinitely strange if *t* never occurs), while the logarithm makes the function much smoother.

A combination of the effect of the occurrence of *t* in particular positions of *p* as explained before, of the anchor texts, and of the TFIDF score $S(p,t)$, determines the relevance value $R(p,t)$ according to the engine's particular policy.

Page relevance with TFIDF metric is one of the two major criteria used for determining the importance of a page in answering user queries. The second criterion is the popularity of a page as a function of its location in the Web graph. The great success that Google experienced since its appearance was undoubtedly connected to the introduction of a new ranking method based on counting the incoming links to each page as a measure of their popularity. The quality of the answers was spectacularly improved over the existing engines.

As previously mentioned, the basic idea was to apply the mathematical concept of a Markov chain to the Web graph to compute the probability of reaching a certain page by a random walk: the more incoming links to a page there are, the greater the probability of visiting it; and so the higher the rank to be assigned to the page for answering users' queries. The proposed algorithm was called Page Rank, and is just one of the ingredients for ranking. According to it, ranking is completely independent of the actual queries. As we also mentioned, around the same time another method called HITS was proposed, also making use of the structure of the Web graph but deciding ranking dynamically as a function of user queries. Both methods are of paramount importance for the development of search engines and will be briefly described here. In fact, in the introduction of this book we have speculated on how to wander through the city of Königsberg in search of art work, following very naively the rules of the two methods.

Figure 9.6 shows the basic structure on Page Rank. In the words of the Google Web site, Page Rank interprets a link from a page to another as a vote by the former to the latter but "*looks at more than the sheer volume of votes, or links a page receives; it also analyzes the page that casts the vote. Vote cast by pages that are themselves "important" weight more heavily ... .*"

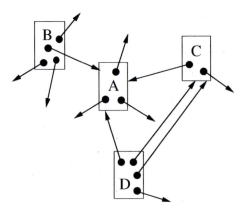

FIGURE 9.6: The basic recursive formula for Page Rank: $R(A) = R(B)/4 + R(C)/2 + R(D)/3$. Note that $R(D)$ is divided by three because two of its four outgoing links point to the same page.

Looking at the figure we can see that a page A with incoming links from B, C, D has its rank $R(A)$ computed as the sum of the ranks of the pages pointing to it, each divided by the number of outgoing links from those pages. For example the contribution of page B to the rank of A is equal to the rank of B proportionally shared by all the pages to which B gives its vote (out-links of B). If more links go from one source to the same destination, as it happens for page D, they count one in the rank computation. So the general formula for the Page Rank $R(p)$ of a page p is as follows:

$$R(p) = \sum_{q \in Q}(R(q)/L(q)) \tag{9.3}$$

where Q is the set of all pages pointing to p, and $L(q)$ is the number of distinct pages pointed by q.

Formula (9.3) is recursive and can be computed starting from an arbitrary assignment of ranks to all the pages (typically, all equal ranks) and applying linear algebra as we shall see below. It can be proved that this is equivalent to making a random tour through the Web graph, taking all outgoing links from any node with equal probability. As a consequence of Markov theory it can also be proved that the initial assignment of rank values does not affect the final result if the number of steps in the random tour (i.e., the number of rank re-calculations) goes to infinity, or in practice is very large. At this point the rank of a page is the probability that the tour terminates there: the implication is that pages with high rank are more likely to be required by users.

Things, however, are not that simple. Users may not always follow the clickable links, and in fact all available statistics show that the average Internet user follows up to three links from a page before getting bored and changing search strategy. So the random tour through the graph foreseen in mathematical terms might actually be shorter than expected. Moreover Page Rank, as defined with formula (9.3), tends to favor older pages because new ones generally have only a few incoming links even if they are actually important, pretty much as it happens with the mechanisms of network growth discussed in Chapter 6. Then a *damping factor* d was added to the formula since the very beginning, accounting for the possibility of jumping from one node to any other node chosen at random. The new ranking formula then becomes:

$$R(p) = d \cdot \sum_{q \in Q}(R(q)/L(q)) + \frac{1-d}{N} \qquad (9.4)$$

where N is the total number of pages in the considered collection, and d is generally taken as 0.85. As a limit, for $d = 1$ formula (9.3) holds, while for $d = 0$ the links have no influence on the ranks and all vertices have equal rank $1/N$.

At this point the computation of Page Rank is a mere application of matrix multiplication as explained in Chapter 6 (Section 6.1). Given the adjacency matrix M of the Web graph, we have seen that any of its powers M^k gives the number paths of length k inside the graph. To take care of damping, we extend the definition of M to a new matrix S whose elements are: $S[i,j] = d \cdot M[i,j] + (1-d)/N$. Numbering the pages from 1 to N, the values of Page Rank can be stored in a vector R where $R[i]$ is the rank of page i. So starting from an initial configuration of values R_0 for R, the computation is iterated as:

$$R_1 = S \times R_0,$$
$$R_2 = S \times R_1, \quad \text{(i.e., } R_2 = S^2 \times R_0\text{)}$$
$$.... \, R_i = S \times R_{i-1} \, ... \quad \text{(i.e., } R_i = S^i \times R_0\text{)} \qquad (9.5)$$

An important point is that we do not actually need to compute the limit rank values, so the chain of computations can be interrupted when the relative standings of the elements of R_i are the same as the ones in R_{i-1}. In other words, it is not necessary to compute the probability of ending in a page A, but just to know whether $R(A)$ is greater or smaller than the rank of the other pages. We will return to this point below.

Compared to Page Rank, Hyperlink Induced Topic Search (HITS) has the computational advantage of working on much smaller arrays and the logical advantage of exploiting the page incoming links in function of the user query, although this last property slows down the phase of query answering.

The idea behind the method is sorting out the pages that are really "authoritative" for a certain query q, from among the many pages with high relevance for the keywords of q and high in-degree. To this end, the set Q of pages containing the keywords are selected together with all the pages pointing to Q or pointed to by Q. The whole set is called the "base" of the query,

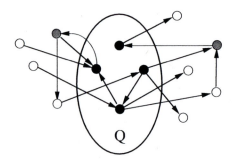

FIGURE 9.7: The "base" of a query q for HITS. Q is the set of pages containing the keywords of q. The base contains Q together with all the pages pointing to Q or pointed to by Q. Grey nodes indicate pages linked to Q in both directions.

see Figure 9.7.[6] The two concepts of authority and hub can be formalized at this point.

Restricting our graph to the base of q, a page p has a non-negative *authority weight* $A(p)$ and a non-negative *hub weight* $H(p)$ mutually reinforcing according to the relations:

$$A(p) = \sum_{s \in S} H(s),$$
$$H(p) = \sum_{t \in T} A(t), \tag{9.6}$$

where S is the set of pages pointing to p, and T is the set of pages pointed to by p. An important authority is a page pointed to by many important hubs; an important hub is a page that points to many important authorities.

Staring with equal values of A and H for all the pages of the base, relations (9.6) are iterated until an equilibrium is reached. The computation is similar to the one of Page Rank, using the much smaller adjacency matrix of the base.

9.3.4   Answering the user queries

The information accumulated by the crawlers, and the page ranking calculated on it, allow search engines to satisfy user needs as well as possible. This essentially means producing a list of answers for any query in order of probably decreasing interest for the user. With the major engines available today this result is often reached with impressive precision, although in other cases the answers provided are scarcely interesting if not at all related to the user intention. Where search engines will be going to improve the quality of

---

[6] In practice each page in Q is allowed to bring into the base a bounded number of pages pointing to it, to keep the size of the base reasonably small. In fact, as already noted there are Web pages in the tail of the scaling distribution that have a huge number of incoming links.

their answers is discussed in the next chapter. Let us see here what happens presently, or at least what is likely to happen because the full ranking strategy is a closely-kept secret of each engine.

First note that search engines accept paid advertising, that essentially implies promoting the rank of a page for a fee. Different marketing policies are used and also the style with which paid pages are shown to the user differ from one engine to another. We will not deal with this aspect of ranking and focus on the methods with which all other pages are treated.

To the basic data structure of Figure 9.3 other tables, or other fields, must be added to keep the ranking scores of the different pages. As seen both TFIDF score and Page Rank can be pre-computed off line and kept in the data structure, thus being ready for answering queries. HITS weights, instead, are computed at query time and may slow down the answering mechanism substantially. So even if in principle HITS prevails over Page Rank in terms of quality of the answers, its applicability is more limited. In addition a wide range of other factors influence the ranking scores, both for improving the quality of answers and for allowing a more rapid response. Most of these factors are undisclosed, but some common criteria certainly apply to all engines.

An important feature is how the three tables D, T, and P are organized, because they are the key for reaching the pages relevant for any query. If binary search or an equivalent tree-search method is used, both D and T are arranged in increasing (alphanumeric) order of the fields docID and term, respectively. Furthermore the docIDs contained in the posting list P for every term must be also stored in increasing order, because this allows to speed up considerably the often required operation of looking for the pages that contain two or more given keywords of the same query. In the example of Figure 9.3, assume that all the documents containing both "beatles" and "submarine" must be found. A search through the tables T and P gives the two lists of docIDs $L_1 = (5,90,...)$ and $L_2 = (20,90,...)$ that may indeed be very long. A naive solution for finding the common elements is searching each element of L_1 into L_2. A much faster method is applying the algorithm DOBLESEARCH presented in Chapter 4 (Section 4.3) that terminates in (optimal) time proportional to the length of the two lists.

An even more interesting feature is assigning the docIDs in order of decreasing ranking score so that the search for the pages containing one or more keywords would encounter the pages with highest score first and could be stopped once the score goes below a certain threshold. This method, however, conflicts with other memory needs too long to be explained here and is only partially used.

Other factors that are likely to be taken care of in all search engines are ranking whole groups of keywords that commonly appear together in the queries; or increasing the rank of pages that are frequently clicked; or even caching the answers for the most common queries.[7] In any case, trying to implement all features that may capture the "intentions" of the user.

9.3.5   The role of distributed and parallel computing

Obviously the tremendous work required to a search engine for crawling and indexing the Web, and for answering queries, would be impossible for a unique even huge computer. The task is then distributed among a large number of reasonably small machines and the computation is largely done in parallel, along the lines discussed in the previous chapter. Again not all is known on how the work is organized, but at least some basic facts have been disclosed.

The largest search engines operate on different centers distributed worldwide, each containing maybe thousands of PCs grouped in clusters, with each cluster dedicated to one of the major search engine functions described above. A first level of parallelism occurs among centers since each of them crawls and index the whole Web, with a major effort to maintain data consistency across the centers. Crawling, indexing, and ranking tasks are done by the computers of the dedicated clusters in a second level of parallelism, under the control of a set of coordinating machines that merge the results. A third level of parallelism takes place inside the computers themselves.

Once a user query is issued, a feverish activity is triggered. The query is delivered to the geographically closest, or the less busy center, and assigned to a specific machine that will provide the answer. The pages in the answer are found through the overall data structure and fetched from the center disks that contain the entire copy of the Web, with a strategy similar to the index look-up phase. The request is issued in parallel to different sections of the archive, each served by a dedicated machine.

Due to their nature, all the operations of a search engine are efficiently done in parallel. Crawling and indexing are shared among independent machines, with a limited amount of work for distributing jobs and collecting results. Even the titanic task of computing Page Rank on the Web adjacency matrix is done with standard techniques of parallel array computation. The most challenging problems have to do with network traffic and bandwidth requirements, but this is a different story.

---

[7]Note that the pages in the engine memory may have been modified or may have even disappeared from the Web after the crawling. The unpleasant sensation of reading a page that the author may have suppressed is sometimes compensated by the interest of still getting something that is no more available.

9.4 Spamming the Web

When the Web became a fundamental means of doing business, it immediately attracted the attention of crooks aiming to unscrupulously increase their financial gains, if not commit fraud by misleading browsers and search engines, and through them their final users.

Techniques aimed at increasing the apparent value of a "target page," exploiting the methods that search engines use for deciding page relevance and/or importance, generally go under the name of Web spamming. This form of deception is the less dangerous for the end user and might even not be considered "dishonesty" unless one can first prove that search engines are in fact "honest."

If page relevance is largely decided through TFIDF metric, as often happens, the obvious spamming method is introducing words frequently demanded by users into the target page p, or into the anchor text of other pages pointing to p, even though these words have nothing to do with the real contents and aims of p. If the spamming goal is to make p relevant for some specific query, just a few specific words must be repeated many times in p to increase its TFIDF score. If instead p must be made relevant for many different queries, many different words must be inserted. In any case a problem is to make such words invisible in p and in the anchor texts. For some time this was done by writing the words in the same color as the page background. Search engines now discard pages with such an explicit feature, so other tricks are used instead, exploiting properties of HTML. While the relevance of a page related to a given set of keywords is given by the presence and the positions of those keywords, the overall rank of the page depends also on its connections in the Web graph. So widely used spamming techniques are aimed at creating artificial link structures, in particular to boost the Page Rank or the authority score of a target page. Recalling how ranking algorithms work, the following two attacks can be readily understood.[8]

To fool the Page Rank algorithm a *spam farm* can be created, consisting of a set of *boosting pages* together with a boosted target page (Figure 9.8). Boosting pages exchange links in two directions with the target. The latter also gets incoming links from a set of *hijacked pages*, that is, innocent pages infiltrated by the spammer for posting the new links, for example the comments section of a blog. All the pages of the farm are then accessible from the Web, and the target page has a high rank due to the links pointing to it, in particular if the spammer can afford to build a big farm and hijack many pages.

To fool HITS a spam farm can be created from scratch without the need to infiltrate existing pages, inserting pointers from the boosting pages to many

---

[8]For some terminology and basic concepts we follow Z. Gyöngyi and H. Garcia-Molina, see the bibliographic notes.

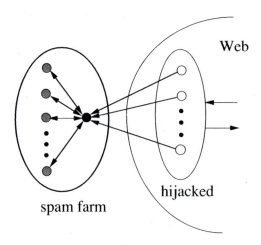

FIGURE 9.8: A spam farm for Page Rank. The black target page gets high Page Rank. Note the links in both directions between the grey boosting pages and the black one.

well known authorities (Figure 9.9). In this way the boosting pages get a high hub score and promote the authority of the target page to which they point.

Clearly things are not that simple. Figures 9.8 and 9.9 indicate two basic structures that get transformed in several ways, both to increase the target score and to make spam farms invisible. In fact all search engines have studied detection tools based on the regularity of the farms that are artificially built, although neither all the techniques used nor the results obtained have been disclosed. Once again the interested reader is invited to follow the ever evolving technical literature.

A combination of spamming methods has been used also for fun. Probably the most famous case of amusing *Google bombing* was directed against the President of the United States. A BBC news article of December 7, 2003 reads:

"Miserable failure" links to Bush.
George W. Bush has been Google bombed. Web users entering the words "miserable failure" into the popular search engine are directed to the biography of the president on the White House website.

Readers may now easily understand how this attack was possible via a farm of pages containing the words "miserable failure" in the anchor text of links pointing to the target page. In that specific case it was reported that as few as thirty two spam pages were sufficient because the two keywords were not particularly common.

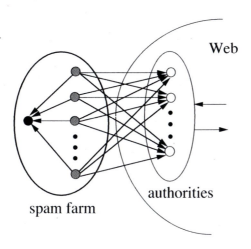

FIGURE 9.9: A spam farm for HITS. Grey pages get high hub score. The black page gets high authority score.

If someone looking for basketball results may merely be a little annoyed at getting a boosted page advertising cameras for sale as the first result from a search engine, other forms of deception are much more serious and possibly dangerous. Fake Web sites abound on the Internet, and are known as *concocted* or *spoof* pages depending on the type of the intended fraud.[9]

Concocted sites offer services or merchandise that will never be performed or sent, collecting money and quickly disappearing. They perform a sort of non-sophisticated attack relying on the naiveness of users (that remains much more widespread than one may expect) and are often detected by software tools on the basis of the domain registration, in particular the host name and country. Regrettably many good sites may look suspicious and be discarded this way, but perhaps it is better not to take risks.

Spoof sites are much more sophisticated and dangerous because their pages are faithful imitations of real sites, typically banks, e.Bay, and the like. The intention is to attract customers of a real site onto the fake one in order to steal personal data or charge money for services that will not be provided. Aside from making use of black lists of URLs at the server side, and white lists at the client side, detection tools perform counterattacks based on the analysis of site links, text similarities, and other technical criteria not publicly disclosed. The most obvious way of attracting a user to a spoof site, called *phishing*, is

---

[9]There is no standard terminology. Here we follow the article of A. Abbasi and H. Chen mentioned in the bibliographic notes.

through an e.mail containing the link, but other complicated methods exist, in particular by infecting servers with fake IP addresses.[10]

Finally recall that search engine crawlers identify themselves to the sites they visit to ask for permission to index the pages. This gives the site the possibility of giving the crawlers a document with a given URL, and serving to the other browsers a different version of the document with the same URL. This is an interesting feature because some elements that need not to be indexed may be excluded beforehand from the crawler version, but it also allows a form of malicious manipulation called *cloaking*. In fact a spammer site can release a clean version of a page to the browsers, so as not to end up on a black list, and a spammed version to the crawlers in order to have the spam terms indexed.

The struggle between villains and search engines continues unceasingly and, as might be expected, without disclosure of the most recent weapons in use on both sides. Some pointers to the open literature given in the bibliographic notes will prove useful if the reader has an interest in the field.

If the reader is instead merely interested at defending her/himself from Web attacks, there is not much that we can suggest except using common sense. In particular:

1. beware of the e.mails from a widow of a foreign general wishing to share several million dollars hidden by her husband before his death;

2. beware of any message written in an improbable language that is clearly the output of an automatic translator, especially if it asks for your personal data;

3. beware of any e.mail from "your" bank (the bank name can be discovered through your browser's history, and in any case banks do not generally communicate via e-mails).

And, a little more seriously, beware of the insidious "social phishing" emails that gain your confidence by apparently being written by your friends, and appear to be genuine because they are full of personal information that can easily be collected from social networks.

Bibliographic notes

Out of a vast technical literature, a classical book on data mining is still a must: Witten, I.H., A. Moffat, and T.C. Bell. 1999 (Second Edition). *Managing Gigabytes*. Morgan Kaufmann, Amsterdam. Among the more recent contributions we suggest: Manning, C.D., P. Raghavan, and H. Schutze. 2008. *Introduction to Information Retrieval*. Cambridge University Press.

---

[10]In 2008 a serious flaw was discovered in the DNS mechanism, that allowed IP addresses of spoof sites to be released through the Internet. The flaw is being fixed in most servers, but new attacks, and new defenses, will certainly follow.

An interesting book on search engines that touches also "political" aspects of the field is: Witten, I.H., M. Gori, and T. Numerico. 2007. *Web Dragons*. Morgan Kaufmann, Amsterdam.

Several survey papers have revised the topics treated in this chapter. In particular many algorithmic problems related to the Internet including an approach to Web caching, with pointers to the scientific literature, are discussed in: López-Ortiz, A. 2005. Algorithmic Foundations of the Internet. *SIGACT News*, Vol. 36, 2, pp. 1-21. ACM Press. A simple and serious introduction to the way search engines work, along the same lines as our Section 9.3, can be found in: Hawking, D. 2006. Web Search Engines: Part 1 and Part 2. *IEEE Computer*, Vol. 39, 6, pp. 86-88, and Vol. 39, 8, pp. 88-90.

The field of Web spamming and attacks is in continuous evolution and any bibliography becomes quickly obsolete. Readers may refer to: Abbasi, A. and H. Chen. 2009. A Comparison of Tools for Detecting Fake Web sites. *IEEE Computer*, Vol. 42, 10, pp. 78-86, and follow the literature, looking, for example, for the contributions of D. Fetterly and M. Jakobsson and for several papers by Z. Gyöngyi and H. Garcia-Molina.

Finally we mention the two articles that first described the mathematical concepts behind Page Rank and HITS, respectively: Brin, S. and L. Page. 1998. Anatomy of a large-scale hypertextual Web search engine. *Computer Networks*, Vol. 30, 1-7, pp. 107-117. The paper first appeared in the Proceedings of the Seventh International World Wide Web Conference. And: Kleinberg, J.M. 1999. Authoritative Sources in Hyperlinked Environment. *Journal of the ACM*, Vol. 46, 5, pp.604-632.

Chapter 10

Epilogue

How humans communicate their knowledge: from Prometheus to Mersenne and Ted Nelson, through Inca messengers, carrier pigeons, and the Web.[1]

After having communicated from time immemorial using sound, mankind discovered writing. According to Aeschylus, Prometheus taught writing to mortals as just one of many useful arts, besides giving them the fire stolen from heaven. As a punishment Zeus chained him to a rock where an eagle perpetually consumed his liver; and there Prometheus lamented his fate:

.... They had neither knowledge of houses built of bricks and turned to face the sun, nor yet of work in wood; but dwelt beneath the ground like swarming ants, in sunless caves. They had no sign either of winter or of flowery spring or of fruitful summer on which they could depend, but managed everything without judgment, until I taught them to discern the risings of the stars and their settings, which are difficult to distinguish. Yes, and numbers, too, chiefest of sciences, I invented for them, and the combining of letters, creative mother of the Muses' arts, with which to hold all things in memory.[2]

So Prometheus gave math and writing to mortals, the latter "to hold all things in memory." For a contemporary Zeus this would have been the most outrageous gift, apt to raise the humans to the level of the gods.

Cultural anthropologists, of course, claim that things went a little differently; but everybody agrees that the birth of writing marked a revolutionary step in the development of mankind. A virtually endless chain of information could be recorded and made available for individual study, and the new medium also favored the development of logical thinking, although Socrates was convinced of the contrary.[3] Dua-Khety, an Egyptian writer of the Middle

---

[1] Finally we present a chapter containing no mathematics. This chapter is a reflection on the power of communication, from the invention of writing to the Internet. This kind of thinking is open to anyone: but perhaps with some solid mathematical knowledge behind us, a more profound reflection may emerge.

[2] Aeschylus, *Prometheus bound*, http://classics.mit.edu/Aeschylus/prometheus.html

[3] Plato, *Phaedrus*.

Kingdom, encouraged his son Pepy to become a scribe explaining that this is the best profession. Extracting freely from the long text:

.... The potter is covered with earth, his clothes being stiff with mud..... The courier goes abroad being fearful of the lions and the Asiatics .... The sandal maker is utterly wretched carrying his tubes of oil .... The washerman launders at the riverbank in the vicinity of the crocodile.... It is to writings that you must set your mind! [4]

The Romans also quickly appreciated the power of writing. The saying "verba volant, scripta manent" (words do fly, but when written remain) attests to the preeminence of writing over talking.

Since then, living without writing has been unthinkable. Alongside communication based on sound, another form based on sight was born, and remained as the only form of data storage external to our brains until 1877, when Thomas Edison invented the phonograph for sound recording and reproduction.

10.1   From mail to telephones

The most ancient form of writing is due to the Sumerian civilization of southern Mesopotamia (now Iraq). Since 3000 B.C. hieroglyphs, and later syllabic symbols, were engraved with a reed on moist clay in a style called cuneiform today, and have survived in hundreds of thousands of texts in archaeological discoveries. A discussion on the fascinating history and development of writing is outside the scope of this book (some specialized literature is mentioned in the bibliographical notes). What counts for us is the possibility of recording information in the form of strings of characters, as explained in Chapter 3, disregarding any physical implementation.

Writing also changed long distance communication. Before its discovery, the delivery of messages to distant places was the job of human messengers. An exemplary system reserved for the use of the royal administration was set up by the Inca in Peru. The delivery of a (probably oral) message was assigned to dispatch riders called *Chasqui* who ran from post to post half a league apart, wearing a white feather hat and announcing their arrival with a trumpet, as reported by the indigenous writer Guaman Poma de Ayala in

---

[4] Adolf Erman (1927) *The literature of the ancient Egyptians*, Matheus and Co. Ltd. London.

1615.[5] Today e.mails directed to the Universidad Nacional de San Antonio Abad del Cusco must be sent @chasqui.unsaac.edu.pe.

With the advent of writing, messages could also be committed to non-human carriers such as pigeons or bottles in the ocean, and the postal service. Some ancient historians wrote about carrier pigeons, but their celebrity came later. The Rothschild family is reported to have built a part of its vast fortune on a private message dispatched by a carrier pigeon announcing Napoleon's defeat in Waterloo (the day was foggy and the optical telegraph did not work, so the banker got the news before his colleagues). Nowadays carrier pigeons are employed for example in Cuba to disseminate emergency messages, a system that has become the object of a careful international study since the December 2004 tsunami in South Asia.

More regular delivery services were provided by postal systems, whose origins are very ancient indeed. The Persians, the Greeks, and the Romans, could all distribute letters throughout their domains in almost the same time experienced today. Sometimes these "letters" were written on strange materials. The peoples of Mesopotamia sent their clay tablets written in cuneiform, and in some cases put them in a clay envelope for delivery. However, the first postal service open to everybody was established by the order of Maximilian I of Augsburg in the 15th century and spread through Europe immediately.

Mail has been considered important in all cultures, showing how networks of relationships have always been fundamentally important. The postal service has also been crucial for the development of culture in general, and of mathematics in particular. Pascal and Fermat interacted regularly by sending letters to each other. A mail exchange between Goldbach and Euler gave rise to the famous Goldbach conjecture on prime numbers that is still open. Descartes described by mail to all his friends the coordinate system that would eventually be named after him. More than anyone, the ingenious friar and mathematician Marin Mersenne acted in the XVII century as the center of one of those intellectual communities called "Republica Literaria" (Republic of Letters) that flourished in Europe and America from the Age of Enlightenment, dispatching and exchanging correspondence with all the greatest scientists of his time.[6] Many other mathematicians discussed their ideas by mail until the middle 1900s, and some of them still do showing a romantic attachment to ancient habits. Computer scientists are generally speaking more modern and exchange e.mails.

Aside from the mail, a broader diffusion of information started with the invention of printing. Newspapers and books spread immediately in great

---

[5]The most complete description of the Inca habits is due to the indigenous writer Felipe Guaman Poma de Ayala in his book: *Nueva Cronica y Buen Gobierno* (New Chronicle and Good Government), hand-written in 1615 and republished today in several editions. There is an ongoing debate on the writing ability of the Incas that the book did not resolve.

[6]Mersenne's collection of letters, published in 1988, consists of seventeen volumes and is regarded as a fundamental reference for studying XVII century mathematics. It may be noted that from Republic of Letters comes the appellation "man of letters" to indicate a scholar (women may have not been accepted...).

quantities as instruments of cultural and political debate, as they remain today. A whole book would be necessary to recount the history of printing.

Another important revolution, similar in some sense to that of the Internet, occurred with the invention of the telegraph, which shrank the world more quickly than ever before, with various consequences in many fields. The first telegraph was optical and consisted of blinking shutters or antennas with whirling arms that could assume several different positions to code letters, numbers, or whole sentences. These gadgets were located on top of towers or high buildings to be recognizable by a telescope, so that, weather conditions permitting (remember the Waterloo pigeon), the message could be sent in a very short time from tower to tower up to the final destination. The system was widely used in Europe. The 1797 edition of the *Encyclopedia Britannica* reports:

The capitals of distant nations might be united by chains of posts, and the settling of those disputes which at present take up months or years might then be accomplished in many hours.

The spread of the electrical telegraph, independently born in the United States and in England, was more painful, due to the skepticism against a non intuitive means of transmission and the consequent difficulty of raising funds. With time some American investors began to understand its enormous potential and started the construction of several telegraphic lines from New York to all the other states of the union. The success was immediate in America and then in Europe, although sending messages was expensive, and stimulated the construction of a transatlantic line between Europe and America, finally completed in 1865 after a number of failures. A few years later a network composed of submarine cables, telegraphic subnetwork and pneumatic post, a sort of Internet of the Victorian age, connected a large part of the world.

The telephone came next and had an immediate and impressive impact.[7] The technology developed fast: operations for connecting a caller and a receiver went from manual to electric, to electronic. Signal amplification allowed the connection of very distant points. Modern transmission techniques allow the sending of several calls on the same cable or radio bridge increasing both the traffic and the income of the telephone companies. But this is recent history and is well known.

---

[7]The early history of the invention of the telephone has the taste of a thriller and was never made really clear in a series of lawsuits. We only say here that the patent of Alexander G. Bell was forensically victorious and commercially decisive, but the Congress of United States stated in 2002: "The life and achievements of Antonio Meucci should be recognized, and his work in the invention of the telephone should be acknowledged."

10.2 Storing information

By definition, a repository of information is a library. The word currently still suggests a collection of traditional books and other written materials, but it is hard to predict how long this will remain the case. Once the political institutions and/or the digital electronics industry manage to provide safe access to all digitized information and its continuous reproduction in the ever developing electronic media, collecting books will probably become the hobby of a small minority. Let us see what has happened up until now.

Almost every country has a National Library, together with a wealth of public or private libraries of good reputation. Competition among them is rare and not that significant; but it was not always so. In ancient times the library of Alexandria (in Egypt today), that has been in the news in recent years in the context of a UNESCO sponsored reconstruction project; and the library of Pergamon (in Turkey today) competed to acquire the most prestigious manuscripts.

It is not surprising that both libraries were founded by two great statesmen: Ptolemy, a general of Alexander the Great, in Alexandria; and Eumenes II of the Attalids in Pergamon. The size of the two libraries was impressive. Shortly after their foundation in the III century B.C. they had over two hundred thousand volumes, although some historians report much higher numbers, and went on growing for two centuries more. But, even more interestingly, they were both associated with big "research centers" to which scholars came from all over the ancient world. Luckily the successors of Ptolemy and Eumenes II continued with the same attitude towards such a wonderful cultural heritage.

Close to the libraries the writing industry flourished, with the Egyptians developing their own papyrus paper, and the Attalids, short of papyrus whose exportation from Egypt was eventually forbidden, inventing parchment, called "pergamena" after the name of the city.

Compared to the innumerable libraries that followed in subsequent centuries up to the present age, it is fair to say that Alexandria and Pergamon remained unsurpassed in terms of organization and aims. The rooms were lined with shelves located in such a way to permit enough ventilation to preserve the manuscripts from humidity. Many scholars were in charge of classification and were competent enough to decide the authenticity or the importance of each manuscript, up to the point that, when Pergamon announced the acquisition of a new oration of Demosthenes, the librarians of Alexandria were able to prove that the original was indeed contained in a manuscript in their possession. Officers were permanently in charge of visiting neighboring countries in search of new acquisitions. And a tremendous cultural and artistic life flourished in the two cities.

Taking a big leap in history we come to the digital libraries of today. Of course it is much simpler to read about Alexandria and Pergamon on

Wikipedia than getting into an ancient boat to visit the two libraries. On the other hand, with the greatest respect to the serious and dedicated people that provide services on the Internet, the librarians of Alexandria and Pergamon that could locate a manuscript for the visitor and give advice on its merits are no longer to be found: not only in the Internet world, but even in a physical library of today.

An intermediate way of merging traditional library preservation with computer technology is the Google Books Library Project, which aims "to digitally scan books from their (the participating libraries) collections so that users worldwide can search them in Google." Many major libraries worldwide have agreed to participate, making it possible to imagine a vast librarian patrimony permanently available online.

10.3   The hypertext revolution

In addition to writing words in phonetic or ideographic alphabets, humans have always used pictures to describe scenes. How images are stored, recognized, and processed in our brain is a highly complicated and hotly debated subject. For computers, however, the recording business is very simple as an image is represented by a binary string coding its "pixels" (very small monochromatic regions in which the image is divided), generally in the form of a mixture of the three basic colors red, green, and blue. Since the standard representation is one byte per color component, or 24 bits for pixel, in principle 2^{24} different color tones can be defined, although far fewer are generally sufficient.

However, the Web also uses a different mode of organizing and making available the information stored, that constitutes a great innovation over the past. Let us recount, then, the hypertext story. We will see how a person of keen imagination and nonstandard perspective truly was ahead of his time.

As we have already said, the Web was originally designed in a huge research center to allow people to work together by exchanging knowledge in a sea of multimedia documents. Besides text, Web pages now include graphics, images, audio and video files. But since the very beginning pages could include links to other pages containing additional information. In a sense a Web page expands in many dimensions through its links, thus becoming part of a "hypertext," although this term is a little pompous. In mathematical terms we are talking of a directed graph, no more: not a string, not an image, but something really new in the world of communications. Now all this appears absolutely normal, but when the system was designed the page structure could have taken many different directions.

The concept of hypertext, however, was not new. It had been proposed long before by Theodor H. (Ted) Nelson. The son of a film director and a

Hollywood actress, Nelson grew up in an age of counterculture. He was aware of the motion effects obtainable in the movies, and the idea of reshaping written documents probably came from there. In 1965 Nelson presented a paper at an important computer conference, advocating the use of computers for text processing, something uncommon in those times. There he suggested the idea of a text that develops through hyperlinks in several dimensions. In his own words:

I knew from my own experiment what can be done for these purposes with card file, notebook, index tabs, edge-punching, file folders, scissors and paste, graphic boards, index-strip frames, Xerox machine and roll-top desk. My intent was not merely to computerize these tasks but to think out (and eventually program) the dream file: the file system that would have every feature a novelist or an absent minded professor could want, holding everything he wanted in just the complicated way he wanted it held, and handling notes and manuscripts in as subtle and complex ways as he wanted them handled.

The term "hypertext" was Nelson's invention. He looked at it as a liquid flowing on a blackboard without definite borders. His dream was the creation of an enormous document formed by all the documents of the universe. When the Web was born some of the dreams of Nelson became reality, but he was not satisfied because the hypertext was tightly linked to the network while he imagined something much more general. Today he is often called a visionary, and his influence in the development of the Web is beyond doubt and probably more profound than is commonly thought.

10.4   Where are we now, and where are we going?

As everybody knows we live in the era of Web 2.0, a buzzword without a precise meaning that generally refers to a set of improvements on what the Web can offer, compared to the original capabilities. We assume that blogs, wikis, social networking and many other ways of linking people through co-operative actions are well known enough not to deserve a specific description. Many other services are now offered on the Web, in particular a wealth of software tools on a free or payment basis, but the global participation aspect is probably the major essence of the present stage of evolution of the Web. Users generate their own texts, express opinions, comment a book or a movie, post personal photos, videos, music. As never before in history, almost everyone has the chance to be part of a universal game.

Enthusiasts see the network as the foundation of a revolution whose roots can be found into the counter-culture of past decades. They see its massive use as a true social movement like environmentalism or punk rock, and its technology as an instrument for personal and social improvement. Like other movements, the Web has provoked worldwide changes causing the rise and fall of major players in diverse activities. Giant industries of telecommunication and computing, and innovative companies adopting e.business models are on the rise, while old fashioned companies, and others suffering from copyright infringement are on the decline. A true revolution or not, the network has played an important role in increasing the participation of people in taking important decisions of overall interest, as demonstrated in the mass crusade against land mines in the 1990s. It can perhaps even push governments towards taking more responsible and transparent actions after the diffusion of confidential documents like those released by Wikileaks.

Unfortunately this freedom allows the spreading of incorrect, perhaps deliberately manipulated pseudo-information, and indeed this happens continuously. A great deal of care must be taken in judging the veracity of news collected from the net, particularly as the human race seems to excel at producing gossip and propagating calumny, as captured perfectly in a famous Italian opera:

Calumny is a little breeze/ a gentle zephyr /which insensibly, subtly, /lightly and sweetly/ commences to whisper.... until, in an unstoppable "crescendo" it .... produces an explosion/ like the outburst of a cannon/ an earthquake, a whirlwind/ which makes the air resound.[8]

Very often people are so captivated by it they do not even wish to find out whether or not the rumor is true.

On more solid grounds, it is important to consider the role that the Internet is having in scientific research. Modern science has its roots in the Age of Enlightenment, but some developments have been possible only with the advent of computers. Now we are entering a new scientific age, where different factors like the immediate access to recently obtained results, the use of algorithms designed and tested by others and embedded in free software, and the availability of enormous sets of data are fostering a new approach to scientific research with the Internet acting as a universal connector. The change in mentality is happening slowly, but some spectacular results have already been achieved, for example in molecular biology with the access to powerful string matching algorithms and the availability of huge data bases of genetic and protein sequences.

So far so good. But very surprisingly the characteristic functionality of the Web, namely mining information through search, is still in its infancy. Although search engines have made huge steps forward over the last few years, their use is essentially limited by a "bag of words" query method where the user merely specifies some keywords; in some cases it is difficult to choose the words in order to get a satisfactory result. Much more might be expected from

---

[8]G. Rossini, *The Barber of Seville.*

such sophisticated giants: at least the possibility of submitting a request that a human could understand and answer properly. An example may be sufficient to clarify this point.

When this book was being written, as an answer to the pair of keywords "berkeley movies" Google gave a link to a well organized table with a listing of movie theaters in the area of Berkeley, California, a summary of their current shows with links to a short description of each movie, their prices, and sundry other useful information. But asking for "Japanese movies at berkeley" the link to the table disappeared and all sort of trash answers came out, starting with the details of a restaurant selling sushi close to a Berkeley movie theater. Notwithstanding the fact that, the same day this experiment was performed, one of the theaters in Berkeley was showing the Akiro Kurosawa's famous epic "The Seven Samurai" as also indicated in the previous table.

It is much more than a problem of page ranking, or the like. One would like to ask: "is there a theater in Berkeley showing some Japanese film?" and expect the search engine to politely reply: "yes, The seven samurai is on show at...," or maybe: "there is no Japanese film on show in Berkeley today, but in Oakland... ." What is quite frustrating is that the search engine "seems to possess" the information we want, but it is impossible to get it out because the engine is incapable of interpreting the query. This is what has led to a wealth of studies on the so called *semantic Web*. And in fact Tim Berners-Lee himself was among the first to advocate this as a necessity.

The problem is difficult because it involves a mixture of artificial intelligence, automatic learning, and natural language processing. The advances necessary may seem a long way off, but research is underway into associating "meaning" to raw data and grouping them into semantic classes, both theoretically and by examining how the users pose their queries or tag their files. The use of social networks for this purpose has been widely adopted, leading to the construction of a taxonomy of popular concepts known as a "folksonomy." We anticipate the emergence of some really interesting new tools in the near future.

In essence, where are we going? The Internet and the Web have created a form of communication unthinkable up until a few decades ago. It attracts strong adjectives like total; immediate; ubiquitous; free — notwithstanding possible requests for payment for a faster connection; democratic — notwithstanding an ever increasing "digital divide"; and uncontrolled and open to everyone - except in some countries. If we exclude more esoteric means of communication such as telepathy, the network may appear to be the ultimate medium.

Of course these are very superficial remarks because the Internet is a very recent achievement, and it is too early to understand where it is going. For sure the tremendous capabilities of computer networks are modifying the way in which people communicate in an extraordinary global anthropological transformation. But in what direction and towards what ends this change will lead us, we cannot claim to understand.

Bibliographic notes

Among technical books, probably the one that best discusses some sociopolitical issues related to the Web is: Witten, I.H., M. Gori, and T. Numerico. *Web Dragons*. 2007. Morgan Kaufmann, Amsterdam.

The story of the hypertext and its inventor is very well narrated in an Italian book that deserves English translation: Castellucci, P. 2009. *Dall'Ipertesto al Web. Storia Culturale dell'Informatica* (From Hypertext to the Web. A Cultural History of Informatics). Laterza, Bari. Reading Ted Nelson's original paper is a real experience, and demonstrates his capability of merging developing concepts of computer science with a traditional knowledge of philosophy and sociology: Nelson, T.H. 1965. A File Structure: for the Complex, the Changing and the Indeterminate. *Proceedings of the 20th ACM National Conference*. ACM Press.

A discussion on the role of the Internet in scientific research can be found in: Hannay, T. 2010. What can the Web do for Science? *IEEE Computer* Vol. 43, 11, pp. 84-87. A survey on the semantic Web can be found in: Mikroyannidis, A. 2007. Toward a Social Semantic Web. *IEEE Computer*, Vol. 40, 11, pp. 113-115. The field, however, is moving continuously.

As far as more ancient history is concerned, a fascinating story of telegraphy, supplied with original papers, comments and curiosities, is found in the book: Standage, T. 1999. *The Victorian Internet*, Berkley Books, New York. Finally, a good book on the origin of writing is: Gaur, A. 1992. *A History of Writing*, The British Library, London.

Finally the evolution and the impact of the Internet on the society of man is discussed almost daily in the press and is worth following. Specialized journals like *IEEE Computer*, cultural magazines like *Wired*, and major newspapers continue to provide reliable information.

Index